IET CONTROL ENGINEERING SERIES 94

Analysis and Design of Reset Control Systems

Other volumes in this series:

Volume 8 **A History of Control Engineering, 1800–1930** S. Bennett
Volume 18 **Applied Control Theory, 2nd Edition** J.R. Leigh
Volume 20 **Design of Modern Control Systems** D.J. Bell, P.A. Cook and N. Munro (Editors)
Volume 28 **Robots and Automated Manufacture** J. Billingsley (Editor)
Volume 33 **Temperature Measurement and Control** J.R. Leigh
Volume 34 **Singular Perturbation Methodology in Control Systems** D.S. Naidu
Volume 35 **Implementation of Self-tuning Controllers** K. Warwick (Editor)
Volume 37 **Industrial Digital Control Systems, 2nd Edition** K. Warwick and D. Rees (Editors)
Volume 39 **Continuous Time Controller Design** R. Balasubramanian
Volume 40 **Deterministic Control of Uncertain Systems** A.S.I. Zinober (Editor)
Volume 41 **Computer Control of Real-time Processes** S. Bennett and G.S. Virk (Editors)
Volume 42 **Digital Signal Processing: Principles, devices and applications** N.B. Jones and J.D.McK. Watson (Editors)
Volume 44 **Knowledge-based Systems for Industrial Control** J. McGhee, M.J. Grimble and A. Mowforth (Editors)
Volume 47 **A History of Control Engineering, 1930–1956** S. Bennett
Volume 49 **Polynomial Methods in Optimal Control and Filtering** K.J. Hunt (Editor)
Volume 50 **Programming Industrial Control Systems Using IEC 1131-3** R.W. Lewis
Volume 51 **Advanced Robotics and Intelligent Machines** J.O. Gray and D.G. Caldwell (Editors)
Volume 52 **Adaptive Prediction and Predictive Control** P.P. Kanjilal
Volume 53 **Neural Network Applications in Control** G.W. Irwin, K. Warwick and K.J. Hunt (Editors)
Volume 54 **Control Engineering Solutions: A practical approach** P. Albertos, R. Strietzel and N. Mort (Editors)
Volume 55 **Genetic Algorithms in Engineering Systems** A.M.S. Zalzala and P.J. Fleming (Editors)
Volume 56 **Symbolic Methods in Control System Analysis and Design** N. Munro (Editor)
Volume 57 **Flight Control Systems** R.W. Pratt (Editor)
Volume 58 **Power-plant Control and Instrumentation: The control of boilers and HRSG systems** D. Lindsley
Volume 59 **Modelling Control Systems Using IEC 61499** R. Lewis
Volume 60 **People in Control: Human factors in control room design** J. Noyes and M. Bransby (Editors)
Volume 61 **Nonlinear Predictive Control: Theory and practice** B. Kouvaritakis and M. Cannon (Editors)
Volume 62 **Active Sound and Vibration Control** M.O. Tokhi and S.M. Veres
Volume 63 **Stepping Motors, 4th Edition** P.P. Acarnley
Volume 64 **Control Theory, 2nd Edition** J.R. Leigh
Volume 65 **Modelling and Parameter Estimation of Dynamic Systems** J.R. Raol, G. Girija and J. Singh
Volume 66 **Variable Structure Systems: From principles to implementation** A. Sabanovic, L. Fridman and S. Spurgeon (Editors)
Volume 67 **Motion Vision: Design of compact motion sensing solution for autonomous systems** J. Kolodko and L. Vlacic
Volume 68 **Flexible Robot Manipulators: Modelling, simulation and control** M.O. Tokhi and A.K.M. Azad (Editors)
Volume 69 **Advances in Unmanned Marine Vehicles** G. Roberts and R. Sutton (Editors)
Volume 70 **Intelligent Control Systems Using Computational Intelligence Techniques** A. Ruano (Editor)
Volume 71 **Advances in Cognitive Systems** S. Nefti and J. Gray (Editors)
Volume 72 **Control Theory: A guided tour, 3rd Edition** J. R. Leigh
Volume 73 **Adaptive Sampling with Mobile WSN** K. Sreenath, M.F. Mysorewala, D.O. Popa and F.L. Lewis
Volume 74 **Eigenstructure Control Algorithms: Applications to aircraft/rotorcraft handling qualities design** S. Srinathkumar
Volume 75 **Advanced Control for Constrained Processes and Systems** F. Garelli, R.J. Mantz and H. De Battista
Volume 76 **Developments in Control Theory towards Glocal Control** L. Qiu, J. Chen, T. Iwasaki and H. Fujioka (Editors)
Volume 77 **Further Advances in Unmanned Marine Vehicles** G.N. Roberts and R. Sutton (Editors)
Volume 78 **Frequency-Domain Control Design for High-Performance Systems** J. O'Brien
Volume 81 **Optimal Adaptive Control and Differential Games by Reinforcement Learning Principles** D. Vrabie, K. Vamvoudakis and F. Lewis
Volume 88 **Distributed Control and Filtering for Industrial Systems** M. Mahmoud
Volume 89 **Control-based Operating System Design** A. Leva *et al.*
Volume 90 **Application of Dimensional Analysis in Systems Modelling and Control Design** P. Balaguer
Volume 91 **An Introduction to Fractional Control** D. Valério and J. Costa
Volume 92 **Handbook of Vehicle Suspension Control Systems** H. Liu, H. Gao and P. Li
Volume 94 **Analysis and Design of Reset Control Systems** Y. Guo, L. Xie and Y. Wang

Analysis and Design of Reset Control Systems

Yuqian Guo, Lihua Xie
and Youyi Wang

The Institution of Engineering and Technology

Published by The Institution of Engineering and Technology, London, United Kingdom

The Institution of Engineering and Technology is registered as a Charity in England & Wales (no. 211014) and Scotland (no. SC038698).

First published 2015

The Institution of Engineering and Technology
Michael Faraday House
Six Hills Way, Stevenage
Herts, SG1 2AY, United Kingdom

www.theiet.org

British Library Cataloguing in Publication Data
A catalogue record for this product is available from the British Library

ISBN 978-1-84919-703-8 (hardback)
ISBN 978-1-84919-704-5 (PDF)

Typeset in India by MPS Limited
Printed in the UK by CPI Group (UK) Ltd, Croydon

Contents

List of figures

List of table

Preface

Reset control is a special kind of hybrid control methods aiming to overcome inherent limitations of linear feedback design. The first investigation of reset control goes back to J.C. Clegg in 1958 for the reset integrator which is called *Clegg integrator* (CI) later. The potential application of reset control in relieving Bode's gain-phase constraints was originally supported by the *describing function* (DF) analysis. The analysis and design of reset control in the time domain have long been considered to be challenging. The first reset control design procedure was proposed by Horowitz and his co-authors in 1975. However, the reset control had gone through a period of silence before attracting renewed attentions in the last 15 years. Some new analysis techniques and design methods for *reset control systems* (RCSs) have been proposed and some of them have been applied to improve performance of some practical control systems.

This book is devoted to the analysis and design of RCSs with emphasis on RCSs with nonzero reset matrices and their applications to *hard disk drive* (HDD) servo systems. Except for the basic theory of reset control, some new ideas for analysis and design of reset control are introduced.

Chapter 1 gives the motivation and basic concepts of RCSs. Some typical design methods including the Horiwitz's design and the PI+CI (proportional-plus-integral control with CI) design are also briefly discussed. In Chapter 2, the DF of reset systems is derived and how the reset matrix affects the frequency domain property of a system is analyzed. In this chapter, the DF is applied to the mid-frequency disturbance rejection of HDD servo systems. Chapter 3 collects some of the recent developments on stability analysis of RCSs. These include quadratic stability of RCSs with stable baseline dynamics and reset times-dependent stability for systems with both stable and unstable baseline dynamics. Passivity of RCSs is also discussed in this chapter. Chapter 4 discusses robust stability of RCSs with uncertainties. Quadratic stability and affine quadratic stability are, respectively, investigated for systems with time-varying and constant uncertainties. Robust stability for RCSs with time-delays is also studied in this chapter. From Chapter 5 to Chapter 7, several different reset schemes are developed. Chapter 5 is about RCSs with discrete-time reset conditions, which can be regarded as a discretized version of the traditional reset mechanism. The discretization provides a different perspective on reset control. A heuristic design procedure aiming to achieve perfect transient response is proposed and applied to track-seeking control of HDD systems. Chapter 6 is about optimal reset control design under fixed reset time instants. Both moving horizon optimization and fixed horizon optimization are discussed and the applications of optimal reset law design to HDD systems and a kind of PZT-positioning stage are, respectively, investigated in this chapter. Chapter 7 provides the main idea of RCSs with conic jump sets and discusses passivity and finite L_2 gain stability under reset control.

The main part of the materials in this book (Chapters 2, 4–6) is based on our research over the past decade. The materials in Chapters 1, 3, and 7 are collections of research results by others. Most of the results in Chapter 3 were originally developed for RCS with zero reset matrices, but have been slightly reformulated to adapt to RCSs with nonzero reset matrices in this book. Chapter 7 briefly introduces some recently developed ideas for reset control.

Yuqian Guo, Lihua Xie, Youyi Wang
26 April, 2015

Acknowledgments

This work was supported in part by the National Natural Science Foundation of China (61074002, 61473315).

Chapter 1
Introduction

1.1 Motivation of reset control

In linear feedback control design, performance specifications can be given in both the time and the frequency domains. In the frequency domain, specifications are usually given in terms of the gain and phase properties of open-loop transfer functions over certain frequency range. For instance, consider a standard linear feedback system depicted in Figure 1.1 where $P(s)$ and $C(s)$ are the plant and the controller, respectively, d_i, d_o, and n are the input disturbance, output disturbance, and measurement noise, respectively, r, y, and e are the reference input, output, and tracking error, respectively. In the frequency domain, the output y is related to r, d_i, d_o, and n through

$$y = S(s)(d_o + P(s)d_i) + (1 - S(s))(r - n) \tag{1.1}$$

where $S(s)$ is the sensitivity function defined by

$$S(s) = \frac{1}{1 + L(s)} \tag{1.2}$$

with $L(s) := P(s)C(s)$, the open-loop transfer function. In general, low-frequency gain of $S(j\omega)$ is expected to be small in order to attenuate the disturbances. This in turn requires that the open-loop gain $|L(j\omega)|$ be large over the low-frequency range. In addition, the open-loop transfer function is required to have sufficient bandwidth for fast response and positioning. On the other hand, in order to reduce the effect of high-frequency noises, the gain of $S(j\omega)$ over the high-frequency range should be large, which requires a small open-loop gain $|L(j\omega)|$ over the same frequency range. These specifications require that the transition of the loop gain from low frequency to high frequency should not be too slow.

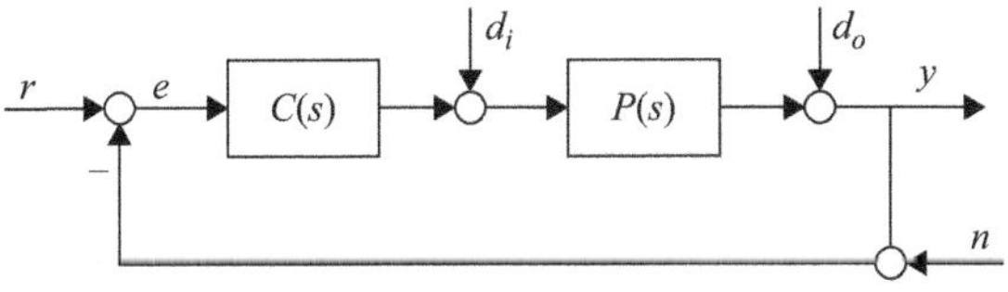

Figure 1.1 Linear feedback control system

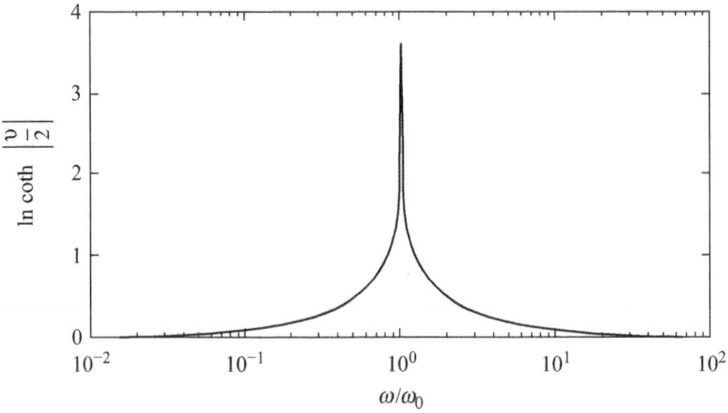

Figure 1.2 The curve of the weighting function ln coth(|v|/2)

However, this requirement often conflicts with the desired phase margin due to *Bode's gain-phase relation*. In order to make this clear, suppose that the open-loop transfer function $L(s)$ is stable and of minimum phase. The Bode's gain-phase relation states that the phase of $L(j\omega)$ is uniquely determined by its gain. Precisely, the phase at frequency ω_0 (rad/s) is given by [1]

$$\angle L(j\omega_0) = \frac{1}{\pi}\int_{-\infty}^{\infty} \frac{d \ln |L(j\omega_0 e^{v})|}{dv} \ln \coth \left|\frac{v}{2}\right| dv \quad \text{(rad)} \tag{1.3}$$

with $v = \ln(\omega/\omega_0)$. The weighting function $\ln \coth(|v|/2) = \ln \left|\frac{\omega+\omega_0}{\omega-\omega_0}\right|$ in the integral is plotted in Figure 1.2. From this figure, one sees that the integration is mainly contributed by the weighted rate of the gain-decreasing near ω_0. Note that

$$\frac{d \ln |L(j\omega_0 e^{v})|}{dv} = \frac{1}{20} \frac{d[20 \lg |L(j\omega)|]}{d \lg(\omega/\omega_0)},$$

thus a rule of thumb states that if the slope of the magnitude curve near ω_0 is $20N$ dB/decade and this slope does not change much over a decade-wide interval near ω_0, then the phase angle at ω_0 can be approximately calculated through [2]

$$\angle L(j\omega_0) \approx \frac{N}{\pi}\int_{-\infty}^{\infty} \ln \coth \left|\frac{v}{2}\right| dv = \frac{N}{\pi}\cdot\frac{\pi^2}{2} = \frac{\pi N}{2} \quad \text{(rad).} \tag{1.4}$$

The above Bode's gain-phase relation indicates that one cannot manipulate the gain and the phase independently and, in order to assure sufficient phase margin, the rate of the gain-decreasing near the crossover frequency ω_c cannot be too fast. In practice, the slope of the open-loop gain at the gain-crossover frequency should be designed to be between -20 and -30 dB/decade in order to assure stability. This tradeoff is explained in Figure 1.3 where the shaded areas represent the bounds defined by the low frequency, high frequency, and phase specifications, respectively. If one wants to design a feedback controller to increase low-frequency gain, broaden the bandwidth, or decrease the high-frequency gain as indicated as the dash line in Figure 1.3, then

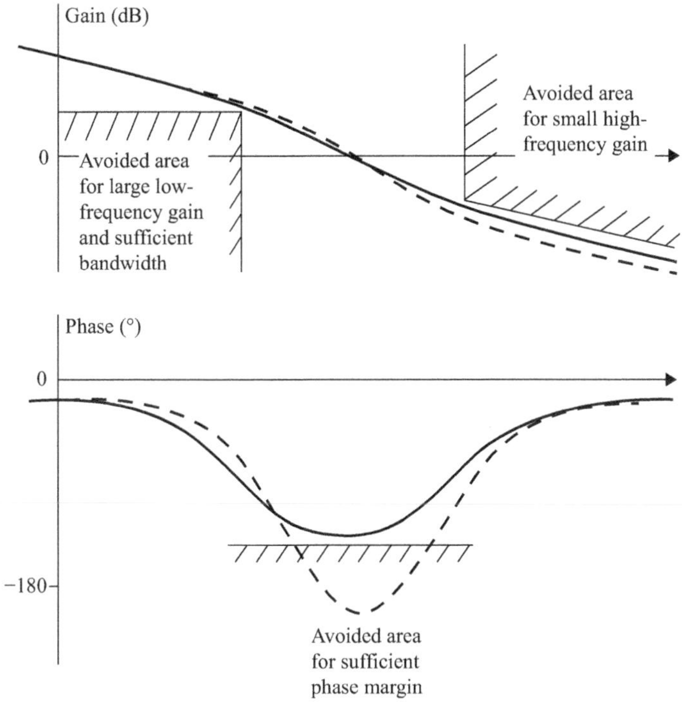

Figure 1.3 Frequency domain specifications

the phase near the crossover frequency would inevitably decrease and even violate the phase margin bound. This tradeoff is unavoidable in the framework of linear feedback control and is called the "cost of feedback" in Reference 3.

The above inherent limitation motivates researchers to adopt nonlinear filters, which are with smaller phase lag but with similar magnitude slope characteristics, to relieve the frequency domain limitations imposed by Bode's gain-phase constraint explained above. Early efforts along this line include the nonlinear feedback proposed by J.B. Lewis [4], the nonlinear gain element by R.E. Kalman [5], and other nonsmooth filters [6, 7]. In recent years, many hybrid control design techniques have been developed to overcome the limitations of the traditional linear feedback design. Such techniques include switching control [8], sliding mode control [9], impulsive control [10], etc.

In this book, we focus on a special kind of hybrid techniques called *reset control* whose original motivation is to overcome the inherent limitation imposed by the Bode's gain-phase constraint. The study of reset control can be traced several decades back. The first reset element is the so-called *Clegg integrator* (CI) proposed by Clegg in 1958 [11]. The CI is described by the impulsive differential equation [11, 12]

$$\begin{cases} \dot{z} = e, & ze > 0 \\ z^+ = 0, & ze \leq 0 \end{cases} \tag{1.5}$$

which consists of a *linear integrator* (LI)

$$\dot{z} = e$$

and a reset mechanism. When the input e and the output z of the integrator have the same sign, then it evolves according to the LI. On the other hand, if the input and the output have opposite signs, then the state is reset to zero. The notation z^+ denotes the state of the integrator after the reset. The condition $ze \leq 0$ is called the *reset condition* which determines when the state of the integrator should be reset. In some recent literature, the triggering condition $ze \leq 0$ is replaced by the zero-crossing of the input, i.e., $e = 0$. In this case, the CI is represented by [13]

$$\begin{cases} \dot{z} = e, & e \neq 0 \\ z^+ = 0, & e = 0. \end{cases} \tag{1.6}$$

These two models of CI are different and lead to different development of reset control theory. However, as basic reset elements, these two models have the same output response to a sinusoidal input. Thus they have the same sinusoidal input *describing function* (DF). Figure 1.4 gives the output responses of the CI and the LI to a sinusoidal input. The DF of the CI is given by [11]

$$G_{CI}(j\omega) = G_{LI}(j\omega)\left(1 + j\frac{4}{\pi}\right) \tag{1.7}$$

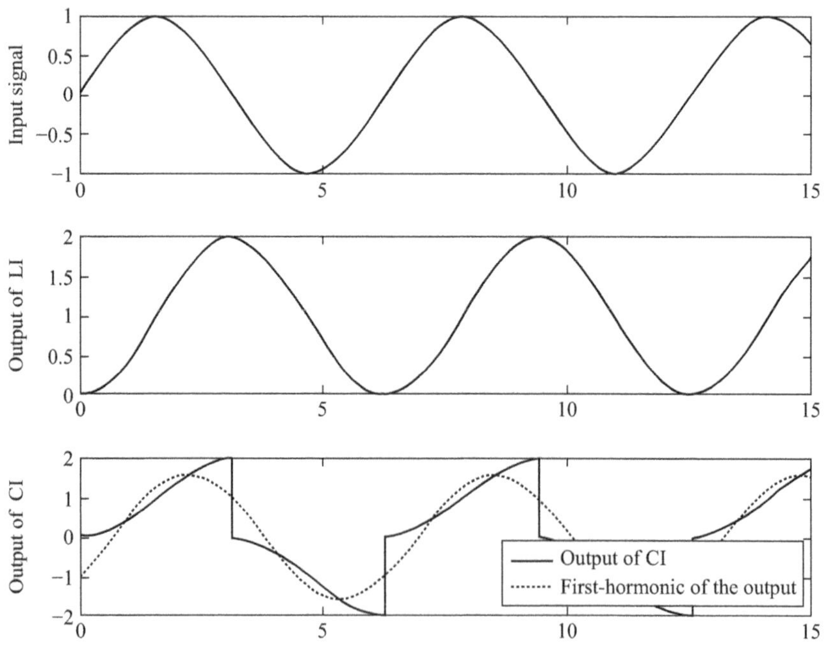

Figure 1.4 Output responses of LI and CI to sinusoidal input e = sin t

where

$$G_{LI}(j\omega) = \frac{1}{j\omega}$$

is the frequency response function of the LI $\dot{z} = e$. The calculation of the sinusoidal input DFs of reset systems will be discussed in detail in Chapter 2. The formula (1.7) shows that the DF of the CI equals the frequency response function of the LI scaled by a complex factor $1 + j(4/\pi)$, which means that the reset action does not change the slope of the logarithmic magnitude-frequency characteristics but causes a phage lead of

$$\phi = \arctan(4/\pi) \approx 51.9^\circ \tag{1.8}$$

at all frequencies. This property suggests that reset elements could be used to achieve required bandwidth with much lower high-frequency gain, which means more rapid gain-decreasing within the mid-frequency range is allowed, without degrading the phase margin specification.

Another basic reset element with more flexibility is the so-called *first-order reset element* (FORE) which was proposed in Reference 3. The state-space representation of a *zero-crossing type FORE* is

$$\begin{cases} \dot{z} = \lambda z + e, & e \neq 0 \\ z^+ = 0, & e = 0 \end{cases}, \quad (\lambda \leq 0) \tag{1.9}$$

whose sinusoidal input DF is

$$G_{FORE}(j\omega) = G_L(j\omega)\,(1 + j\Theta(\omega)) \tag{1.10}$$

where

$$G_L(j\omega) = \frac{1}{j\omega - \lambda}$$

is the frequency response function of the first-order linear element $\dot{z} = \lambda z + e$ and

$$\Theta(\omega) = \frac{2}{\pi}\frac{1 + e^{\pi\lambda/\omega}}{1 + (\lambda/\omega)^2} > 0.$$

The Bode plots of the FORE and the corresponding first-order linear element are compared in Figure 1.5. Different from the CI, a FORE behaves like a linear element over low-frequency range but like a CI over high-frequency range. This property provides more useful design flexibility [3]. The DFs of the CI and the FORE have been applied to many practical systems showing benefits of reset control. See for instance References 14 and 15.

The benefit of reset control can also be shown in the time domain. In practice, integral control is usually used to remove steady-state error. But the side effect is that it may cause large overshoot and prolong the settling time. Theoretically, differential control can be used to reduce the overshoot caused by integration without degrading the rise time specification, but in practice, differential control may make the system sensitive to sensor noise. Replacing the integrator by the CI can help to achieve better tradeoff between rise time and overshoot. This benefit was recognized in the early

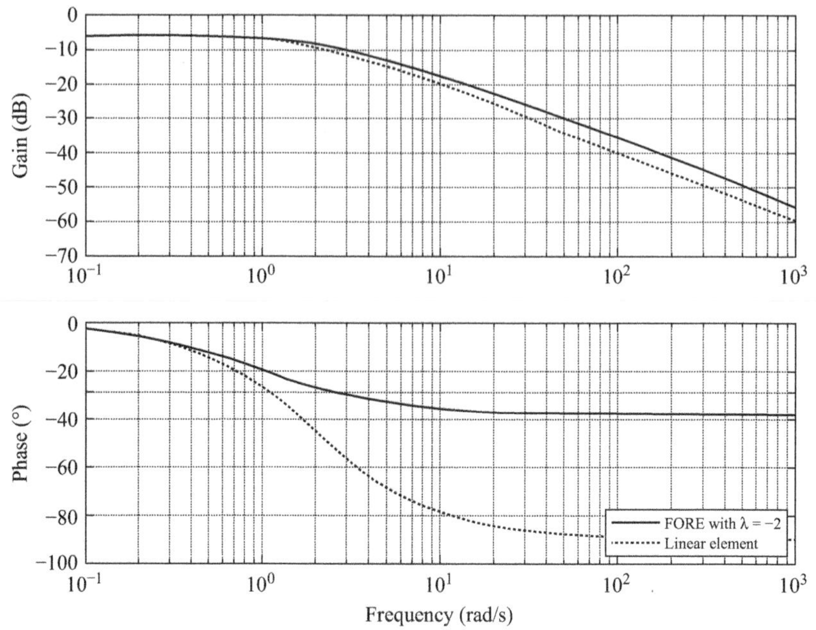

Figure 1.5 Bode plots of FORE and first-order linear element

literature regarding reset control and the first quantitative design procedure around the CI was proposed in Reference 16. The control system considered in Reference 16 is depicted in Figure 1.6. In order to explain the effect of the CI, let us first derive an equivalent representation of the CI.

Suppose that $0 \leq t_1 < t_2 < \cdots < t_n$ are the time instants of the zero-crossings of the input of the CI (1.6). Thus the output of the CI $z(t)$ with $t_0 = 0$ is

$$
\begin{aligned}
z(t) &= \int_{t_l}^{t} e(s)ds, \quad t \in [t_l, t_{l+1}) \\
&= \int_{0}^{t} e(s)ds - \sum_{t_k \leq t} \int_{t_{k-1}}^{t_k} e(s)ds \\
&= \int_{0}^{t} \left[e(s) - \sum_{k=1}^{n} A_k \delta(t - t_k) \right] ds
\end{aligned}
$$

where A_k is the signed area

$$A_k = \int_{t_{k-1}}^{t_k} e(s)ds.$$

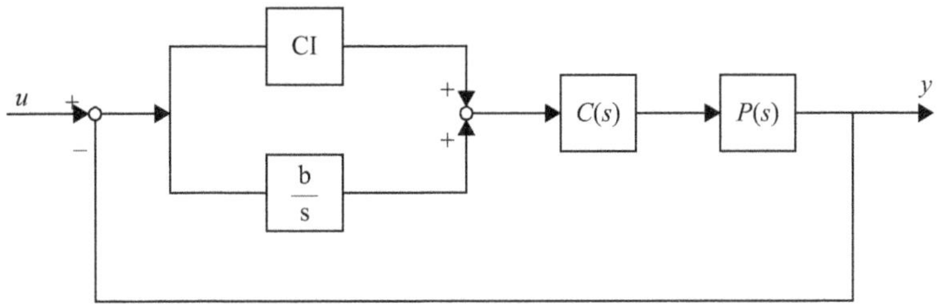

Figure 1.6 The reset control system (RCS) with a CI

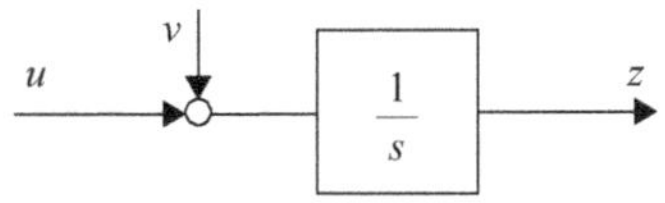

Figure 1.7 CI equivalence

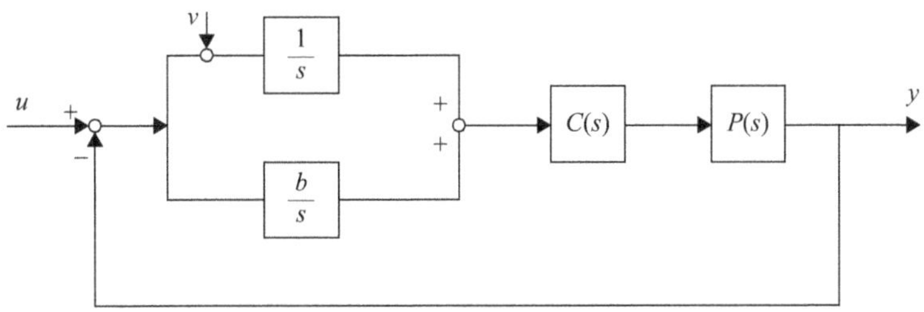

Figure 1.8 Replacement of the CI with its equivalence

Here $\delta(t)$ represents the *unit impulse*. This indicates that the CI can be equivalently replaced by a LI with an additional input of a sequence of impulses

$$v(t) = -\sum_{k=1}^{n} A_k \delta(t - t_k)$$

as depicted in Figure 1.7. Replacing the CI in Figure 1.6 by the equivalence, we get the structure depicted in Figure 1.8. Suppose that the transfer function from u to y (i.e., the closed-loop transfer function of the system with the CI replaced by the LI) is $T(s)$. Then the transfer function from v to y is $\frac{1}{1+b}T(s)$. Thus if the step response and the unit impulse response of the linear system $T(s)$ without reset are, respectively, denoted by $y_L(t)$ and $y_I(t)$, then the output of the system with CI is

$$y(t) = y_L(t) - \frac{1}{1+b}\sum_{k=1}^{n} A_k y_I(t - t_k).$$

Suppose that the step response of the linear system $T(s)$ exhibits overshoot as depicted in Figure 1.9 and t_1 and t_2 are the first two reset time instants. Then before

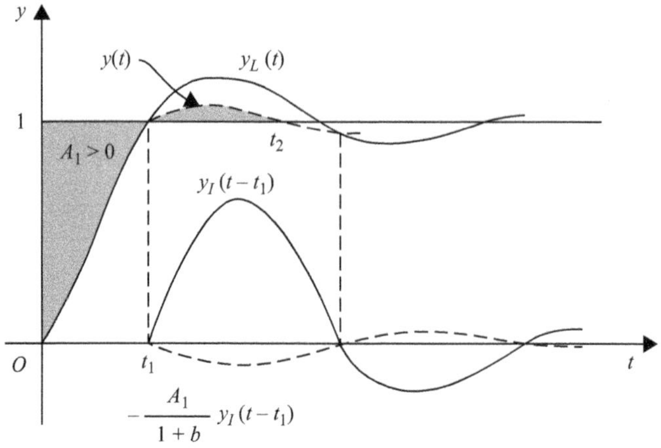

Figure 1.9 Reducing overshoot through reset

the first reset, the output response of the system with CI is the same as the linear system. During the interval $[t_1, t_2]$, there holds that

$$y(t) = y_L(t) - \frac{1}{1+b} A_1 y_I(t - t_1).$$

Note that $A_1 > 0$, thus the reset action at t_1 plays the role of reducing the first peak. If the underlying linear system $T(s)$ is of the second order with damping ratio $\zeta \in [0, 1)$, then the precise value of the overshoot can be calculated as [16, 17]

$$M_R(\zeta, b) = \frac{1}{1+b}\left\{b - 2\zeta \exp\left[\frac{\zeta}{\sqrt{1-\zeta^2}}(\pi - \arccos\zeta)\right]\right\} M_L \tag{1.11}$$

where M_L is the overshoot of the step response of the linear system $T(s)$ given by

$$M_L = \exp\left[-\frac{\pi\zeta}{\sqrt{1-\zeta^2}}\right].$$

It is easy to check that if

$$b \geq 2\zeta \exp\left[\frac{\zeta}{\sqrt{1-\zeta^2}}(\pi - \arccos\zeta)\right], \tag{1.12}$$

then

$$0 \leq M_R(\zeta, b) < M_L.$$

If b is chosen to be

$$0 < b < 2\zeta \exp\left[\frac{\zeta}{\sqrt{1-\zeta^2}}(\pi - \arccos\zeta)\right],$$

then

$$M_R(\zeta, b) < 0,$$

i.e., undershoot occurs. The above analysis indicates that if the parameter b is appropriately chosen, the replacement of the LI by the CI can reduce or even remove the overshoot without degrading the rise time specification.

One interesting question is that if the performance achieved by the reset control can also be achieved by a linear compensator. The following example, which is constructed in Reference 18, gives a negative answer. Consider a linear control system depicted in Figure 1.1 with $d_i = 0$, $d_o = 0$, and $n = 0$ where the plant $P(s)$ contains an integrator and $C(s)$ is stabilizing. Suppose that the reference input is the unit step, i.e., $r(t) = 1(t)$. Define

$$z(t) = \int_0^t e(s)ds.$$

Then the Laplace transform of $z(t)$ is

$$Z(s) := \mathcal{L}(z(t)) = \frac{1}{s}\mathcal{L}(e(t)) = \frac{1}{s}E(s) = \frac{1}{s^2}\frac{1}{1 + P(s)C(s)}.$$

Thus by the final value theorem of Laplace transform, there holds

$$\int_0^\infty e(s)ds = \lim_{s\to 0} sZ(s) = \frac{1}{\lim_{s\to 0} sP(s)C(s)} := \frac{1}{K_v}.$$

Consider the rise time defined by

$$t_r := \sup_T \left\{T : y(t) \le \frac{t}{T},\ t \in [0, T]\right\}. \tag{1.13}$$

There holds

$$\begin{aligned}\frac{1}{K_v} &= \int_0^\infty e(s)ds \\ &\ge \int_0^{t_r} \left(1 - \frac{s}{t_r}\right) ds + \int_{t_r}^\infty e(s)ds \\ &= \frac{t_r}{2} + \int_{t_r}^\infty e(s)ds.\end{aligned}$$

If $t_r > 2/K_v$, then

$$\int_{t_r}^\infty e(s)ds < 0$$

which means that the unit step response overshoots. Keep this in mind and consider the special case that $P(s) = 1/s$. Suppose that we aim to design a controller such that the system is stable and satisfies:

1. the steady-state error of the unit ramp response is not greater than 1;
2. the rise time t_r of the unit step response is greater than 2 s;
3. the step response does not overshoot.

The requirements (1) and (2) mean that the velocity error constant $K_v \ge 1$ and the rise time $t_r > 2 \ge 2/K_v$. Thus according to the previous analysis, if a linear control is used,

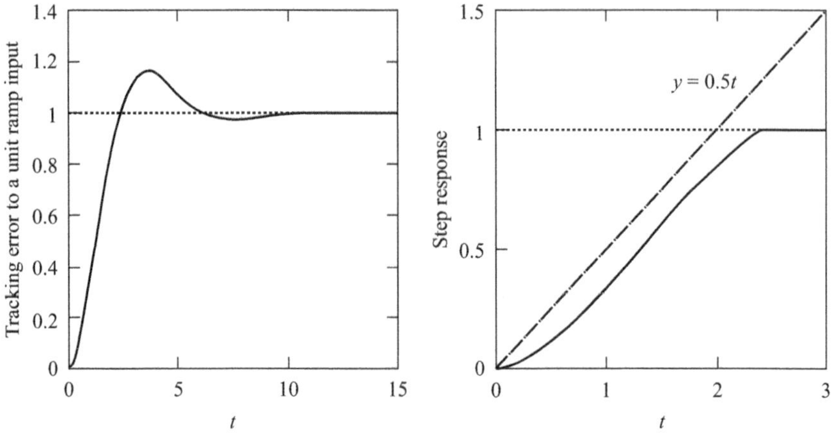

Figure 1.10 Responses of the RCS to unit ramp and step inputs, respectively

then the step response will exhibit an overshoot. This contradicts the requirement (3). Thus there exists no stabilizing linear controller which meets all these specifications. However, these specifications can be satisfied by replacing $C(s)$ with the FORE (1.9) with $\lambda = -1$. See Figure 1.10 which shows that, with this reset controller, the steady-state error of the unit ramp response is 1 and the rise time is approximately 2.42 s. In addition, the unit step response is deadbeat, thus no overshoot occurs.

Though reset control has many benefits over linear control, it is worth pointing out that these benefits do not come from any blind resetting but from an appropriate interaction between the reset mechanism and the underlying linear system. Resetting might deteriorate the system performance or even destabilize the system if the underlying linear controller is not appropriately designed.

1.2 Basic concepts of RCSs

1.2.1 Preliminaries and problem setup

Consider the *single-input single-output* (SISO) linear time-invariant plant

$$\begin{cases} \dot{x}_p = A_p x_p + B_p u_p \\ y = C_p x_p, \end{cases} \tag{1.14}$$

where $x_p \in \mathbb{R}^{n_p}$, $u_p \in \mathbb{R}^1$, $y \in \mathbb{R}^1$, and A_p, B_p, and C_p are constant matrices with proper dimensions. A conventional reset controller can be modeled by the impulsive differential equation

$$\begin{cases} \dot{x}_c = A_c x_c + B_c e, & e \neq 0 \\ x_c^+ = \rho_c(x_c, r), & e = 0 \\ u = C_c x_c + D_c e \end{cases} \tag{1.15}$$

where $x_c \in \mathbb{R}^{n_c}$ is the controller state, $e = r - y$ is the tracking error, and r is the reference input, A_c, B_c, C_c, and D_c are constant matrices, ρ_c is the *reset mapping*, and x_c^+ represents the state of the controller after reset, i.e.,

$$x_c^+(t) = \lim_{s \to t+0} x_c(s).$$

Combining the plant (1.14) and the reset controller (1.15), we have the closed-loop system

$$\begin{cases} \dot{x} = Ax + Br, & r - Cx \neq 0 \\ x^+ = \rho(x, r), & r - Cx = 0 \end{cases} \tag{1.16}$$

with $x = [x_p^T, x_c^T]^T$ and

$$A = \begin{pmatrix} A_p - B_p D_c C_p & B_p C_c \\ -B_c C_p & A_c \end{pmatrix}, \quad B = \begin{pmatrix} B_p D_c \\ B_c \end{pmatrix}$$

$$C = \begin{pmatrix} C_p & 0 \end{pmatrix}, \quad \rho(x, r) = \begin{pmatrix} x_p \\ \rho_c(x_c, r) \end{pmatrix}.$$

The system

$$\dot{x} = Ax + Br \tag{1.17}$$

is called the *baseline system* of the reset system (1.16). For set-point regulation, the reference input r is a constant input. In this case, a state $\bar{x}_r$ is called as an *equilibrium point* of reset system (1.16) if

$$\begin{cases} A\bar{x}_r + Br = 0 \\ C\bar{x}_r - r = 0 \\ \rho(\bar{x}_r, r) - \bar{x}_r = 0. \end{cases} \tag{1.18}$$

Any state satisfying the first two equations of (1.18) is an equilibrium point of the baseline system, and any point satisfying the last equation of (1.18) is an equilibrium point of the reset mapping. Equation (1.18) actually imposes a constraint on the reset mapping, i.e., the reset action should not destroy the equilibrium point of the baseline system. Precisely, suppose that $\bar{x}_r = [\bar{x}_{pr}^T, \bar{x}_{cr}^T]^T$ is the equilibrium point of the baseline system which is to be stabilized. Then the reset mapping should be chosen such that $\bar{x}_r$ is also an equilibrium point of the reset mapping. In the conventional reset control, the reset mapping is of the form

$$\rho_c(x_c, r) = \bar{x}_{cr} + R_c(x_c - \bar{x}_{cr}), \tag{1.19}$$

where the matrix R_c is called the *reset matrix* which is of the form

$$R_c = \begin{pmatrix} I_{n_c - n_\rho} & 0_{(n_c - n_\rho) \times n_\rho} \\ 0_{n_\rho \times (n_c - n_\rho)} & 0_{n_\rho \times n_\rho} \end{pmatrix} \tag{1.20}$$

with n_ρ an integer satisfying $0 \leq n_\rho \leq n_c$. I_m and $0_{m \times n}$ represent the identity matrix with dimension m and the $m \times n$ zero matrix, respectively. This structure means that whenever the tracking error crosses zero, part of the controller state is reset to its steady-state value.

Define

$$\xi = x - \bar{x}_r,$$

then the closed-loop RCS (1.16) becomes

$$\begin{cases} \dot{\xi} = A\xi, & C\xi \neq 0 \\ \xi^+ = R\xi, & C\xi = 0 \end{cases} \tag{1.21}$$

with

$$R = \begin{pmatrix} I_{n_p} & 0 \\ 0 & R_c \end{pmatrix}.$$

In RCSs, the FORE is given by

$$\begin{cases} \dot{z} = \lambda z + e, & e \neq 0 \\ z^+ = z_r, & e = 0 \end{cases} \tag{1.22}$$

where z is a scalar variable, z_r represents the steady-state value of the FORE, and μ is a scalar constant. If $\lambda \neq 0$ and the steady-state error is zero, then $z_r = 0$. Thus a FORE is generally defined as

$$\begin{cases} \dot{z} = \lambda z + e, & e \neq 0 \\ z^+ = 0, & e = 0. \end{cases} \tag{1.23}$$

If $\lambda = 0$, the FORE degenerates to the CI in which case the steady-state value of the integrator is not necessarily zero. Traditionally, the CI is defined as

$$\begin{cases} \dot{z} = e, & e \neq 0 \\ z^+ = 0, & e = 0 \end{cases} \tag{1.24}$$

which should be understood as the case that the equilibrium point has been moved to the origin through linear transformation.

Denote

$$\mathcal{J} = \ker C := \{\xi \in \mathbb{R}^{n_p+n_c} \mid C\xi = 0\} \tag{1.25}$$

which is called the *jump set* of the RCS (1.21). With this definition, the closed-loop RCS can be alternatively represented by

$$\begin{cases} \dot{\xi} = A\xi, & \xi \notin \mathcal{S} \\ \xi^+ = R\xi, & \xi \in \mathcal{J} \end{cases} \tag{1.26}$$

where $\xi \in \mathbb{R}^n$ with $n := n_p + n_c$. $\mathcal{S} = \mathcal{J}^c$ represents the complementary set of $\mathcal{J}$ which is called the *flow set*. The reset system (1.26) with $\mathcal{J}$ defined in (1.25) is called the *zero-crossing type reset system* since the reset action is triggered by the zero-crossing of the tracking error. The jump set is a hyperplane as depicted in Figure 1.11(a). In general, for single output systems, if Rank $C = 1$, then we have

$$\dim(\mathcal{J}) = \dim(\ker C) = n - 1.$$

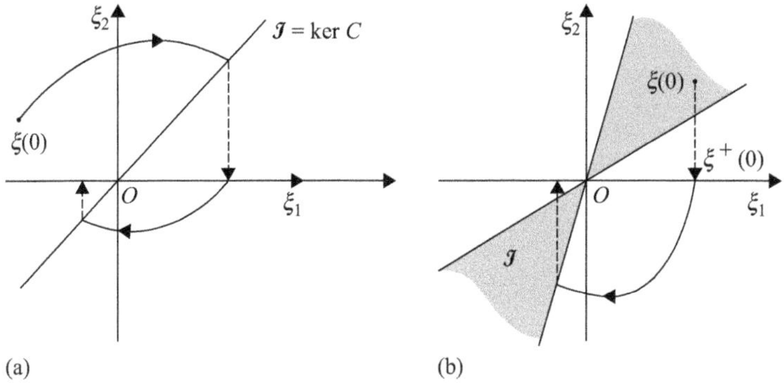

Figure 1.11 Two different triggering conditions (a) zero-crossing type; (b) sector type

The formula (1.26) can be used to model a broader class of RCSs for more general definitions of flow and jump sets. For instance, in Reference 12, FORE is modeled as

$$\begin{cases} \dot{z} = -\lambda z + e, & ez \geq 0 \\ z^+ = 0, & ez \leq 0, \end{cases} \tag{1.27}$$

where the reset action is triggered whenever its input and output have opposite signs. If the control loop includes such a reset element, the resulting reset system will be of the form of (1.26) with *conic jump sets*, i.e.,

$$\mathcal{S} = \left\{ \xi \in \mathbb{R}^n \mid \xi^T S \xi \geq 0 \right\}, \tag{1.28}$$

$$\mathcal{J} = \left\{ \xi \in \mathbb{R}^n \mid \xi^T J \xi \geq 0 \right\}, \tag{1.29}$$

where S and J are symmetric matrices. In this case, the flow and jump sets are even allowed to be partially overlapped, leading to nonunique solutions.

1.2.2 Solutions to RCSs

Consider reset system (1.26) with $\mathcal{S} = \mathcal{J}^c$, the complementary set of $\mathcal{J}$. For any $\xi_0 \in \mathbb{R}^n$, a function $\xi(t; \xi_0)$ of time t is said to be a solution to RCS (1.26) starting from the initial state ξ_0 if there is a countable or finite subset $\mathscr{I}(\xi_0) \subset \mathbb{R}^+$ such that:

1. $\xi(t; \xi_0)$ is left-continuous in time t for $t > 0$ with $\xi(0; \xi_0) = \xi_0$;
2. $\xi(t; \xi_0)$ is differentiable in t and satisfies

   $$\dot{\xi}(t; \xi_0) = A\xi(t; \xi_0), \quad \forall t \in \mathbb{R}^+ \setminus \mathscr{I}(\xi_0);$$

3. For all $t \in \mathscr{I}(\xi_0)$, there hold that $\xi(t; \xi_0) \in \mathcal{J}$ and

   $$x(t^+; \xi_0) - Rx(t; \xi_0).$$

In the above definition, $\mathscr{I}(\xi_0)$ is actually the set of reset time instants of the solution $\xi(t; \xi_0)$. Since it is required to be countable or finite, one can arrange the

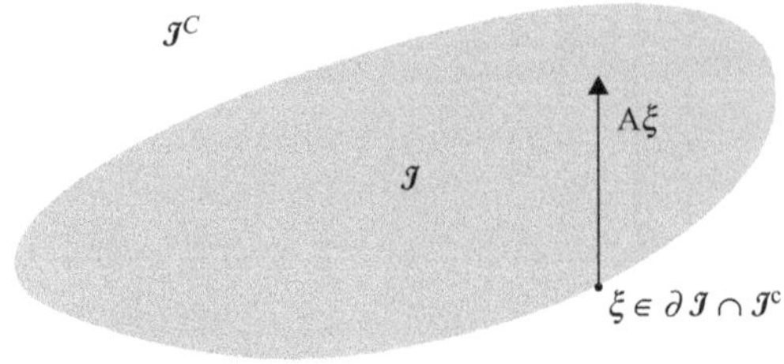

Figure 1.12 Deadlock

reset time instants in an increasing order and define $\tau_k(\xi_0)$ the kth reset time instant of the solution $\xi(t; \xi_0)$. The *reset interval* is defined as

$$\Delta_k(\xi_0) := \tau_k(\xi_0) - \tau_{k-1}(\xi_0)$$

with $\tau_0(\xi_0) := 0$.

One important question is that under what conditions, the reset system exists a well-defined solution for any initial condition ξ_0. Note that the state-dependent reset mechanism defined above may lead to complex phenomena including *deadlock*, *beating*, and *Zenoness* which destroy the existence of solutions. Deadlock happens when the system can neither evolve continuously nor jump discretely. Figure 1.12 gives a situation of deadlock where $\xi \in \partial\mathcal{J} \cap \mathcal{J}^c$ and the vector $A\xi$ directs toward the inner of $\mathcal{J}$. Here, $\partial\mathcal{J}$ represents the boundary of $\mathcal{J}$. Since $\xi \notin \mathcal{J}$, jump is impossible. However, the trajectory of the baseline system starting from ξ will go inside the jump set since the vector field points toward the inner of $\mathcal{J}$, which is not allowed for the RCS. In order to avoid deadlock, the RCS should satisfy the following assumption.

Assumption 1.1. *For any $\xi \in \partial\mathcal{J} \cap \mathcal{J}^c$, there exists a positive number ϵ such that*

$$\exp(At)\xi \in \mathcal{J}^c, \quad \forall t \in (0, \epsilon). \tag{1.30}$$

Beating occurs when there are multiple resettings at a single reset time instants, which is not allowed by the definition of solution. The following assumption is imposed to avoid beating.

Assumption 1.2. *The jump set and its image under the reset mapping are disjoint, i.e.,*

$$\mathcal{J} \cap R(\mathcal{J}) = \emptyset. \tag{1.31}$$

For a given RCS, if $\mathcal{J} \cap R(\mathcal{J}) \neq \emptyset$, one can re-define the jump set by removing $R(\mathcal{J})$ from $\mathcal{J}$, i.e.,

$$\tilde{\mathcal{J}} := \mathcal{J} \setminus R(\mathcal{J}), \tag{1.32}$$

to avoid beating.

Zenoness is a phenomenon where there are infinite number of resettings within a compact time interval in which case the solution is only well-defined within a finite time interval. In order to avoid Zenoness, the following assumption is made.

Assumption 1.3. *For any $\xi_0 \in \mathbb{R}^n$, there is a positive number $\mu(\xi_0) > 0$ such that*

$$\Delta_k(\xi_0) \geq \mu(\xi_0), \quad \forall k.$$

Note that if Assumption 1.3 holds, then it implicitly implies that deadlock and beating would not occur and for any initial state ξ_0, there is a unique solution $\xi(t;\xi_0)$ to RCS (1.26) which can be represented as

$$\xi(t;\xi_0) = W(t,\xi_0)\xi_0 \tag{1.33}$$

where $W(t,\xi_0)$ is the *state transition matrix* given by

$$W(t,\xi_0) = \exp\left(A(t-\tau_k(\xi_0))\right)R\exp\left(A\Delta_k(\xi_0)\right)\cdots R\exp\left(A\Delta_0(\xi_0)\right) \tag{1.34}$$

for $t \in (\tau_k(\xi_0), \tau_{k+1}(\xi_0)]$.

In general, Assumption 1.3 is difficult to check since it depends on the reset time instants directly. In practice, the existence of solutions can be assured by *time-regularization*. For instance, with time-regularization, the reset controller (1.15) becomes

$$\begin{cases} \dot{x}_c = A_c x_c + B_c e, & (e \neq 0) \vee (\tau \leq \Delta_m) \\ x_c^+ = \rho_c(x_c, r), & (e = 0) \wedge (\tau > \Delta_m) \\ u = C_c x_c + D_c e \end{cases} \tag{1.35}$$

where Δ_m is a pre-specified positive number, $\vee$ and $\wedge$ represent the logical OR and AND, respectively, and τ is a variable characterizing the time spent since the latest reset which is governed by the impulsive differential equation

$$\begin{cases} \dot{\tau} = 1, & (e \neq 0) \vee (\tau \leq \Delta_m) \\ \tau^+ = 0, & (e = 0) \wedge (\tau > \Delta_m). \end{cases} \tag{1.36}$$

The structure of RCS is depicted in Figure 1.13. The closed loop becomes

$$\begin{cases} \dot{\xi} = A\xi, & (\xi \notin \mathcal{J}^c) \vee (\tau \leq \Delta_m) \\ \xi^+ = R\xi, & (\xi \in \mathcal{J}) \wedge (\tau > \Delta_m). \end{cases} \tag{1.37}$$

With time-regularization, two successive reset time instants are separated by at least Δ_m units of time. Thus deadlock, beating or Zenoness would not happen and for any initial state, a unique well-defined solution always exists.

1.2.3 RCSs with discrete-time reset conditions

In practical computer-based implementations, reset controllers have to be discretized. For instance, the simplest discrete-time counterpart of the zero-crossing condition

$$e(t) = 0 \tag{1.38}$$

in reset controller (1.15) is

$$e(kT_s)e((k-1)T_s) < 0 \tag{1.39}$$

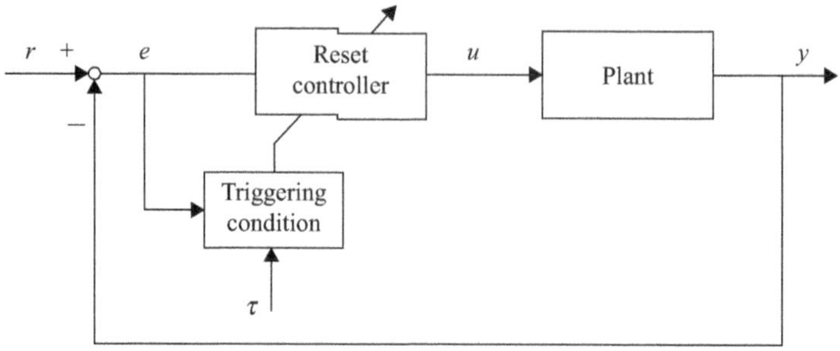

Figure 1.13 An RCS with time-regularization

where T_s is the sampling period. In addition, reset actions occur only at sampling time instants. Thus the discrete-time form of the reset controller (1.15) can be represented by

$$\begin{cases} \dot{x}_c = A_c x_c + B_c e, & t \in ((k-1)T_s, kT_s] \\ x_c(kT_s^+) = x_c(kT_s), & e_k e_{k-1} > 0 \\ x_c(kT_s^+) = \rho_c(x_c(kT_s), r), & e_k e_{k-1} \le 0 \\ u = C_c x_c + D_c e \end{cases} \tag{1.40}$$

where $e_k := e(kT_s)$. A more general discrete-time reset controller considered in this book is of the form

$$\begin{cases} \dot{x}_c = A_c x_c + B_c e, & t \in ((k-1)T_s, kT_s] \\ x_c(kT^+) = x_c(kT), & \phi(e_k, e_{k-1}) > 0 \\ x_c(kT_s^+) = \rho_c(x_c(kT_s), r), & \phi(e_k, e_{k-1}) \le 0 \\ u = C_c x_c + D_c e. \end{cases} \tag{1.41}$$

where

$$\phi : \mathbb{R}^2 \to \mathbb{R}$$

is called the *triggering function* which characterizes the reset condition. This form of triggering condition is capable of describing more general cases. A useful kind of triggering functions is given by the quadratic form

$$\phi(\eta) = \eta^T S \eta, \quad \eta \in \mathbb{R}^2 \tag{1.42}$$

where S is a symmetric matrix. For instance, the triggering function with

$$S = \begin{bmatrix} 0 & (\Delta/T_s + 1)/2 \\ (\Delta/T_s + 1)/2 & -\Delta/T_s \end{bmatrix} \tag{1.43}$$

defines a reset law under which reset time instants are ahead of zero-crossings, where $\Delta > 0$ is a nonnegative constant roughly characterizing the time for which the reset action is ahead of the zero-crossing. Actually, in this case,

$$\phi(e_k, e_{k-1}) = e_{k-1} \left[\frac{e_k - e_{k-1}}{T_s} \Delta + e_k \right] \tag{1.44}$$

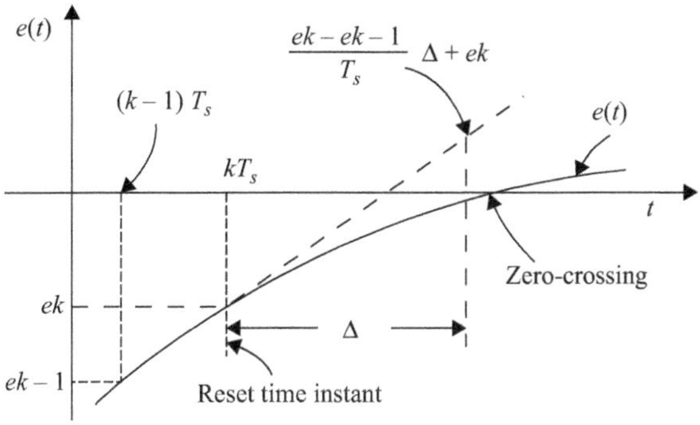

Figure 1.14 Discrete-time reset law defined by the triggering function given in (1.44)

where

$$\tilde{e}(kT_s + \Delta) = \frac{e_k - e_{k-1}}{T_s}\Delta + e_k$$

is a simple linear predictor of the tracking error at $t = kT_s + \Delta$, as depicted in Figure 1.14.

Another kind of RCSs with discrete-time reset conditions is the RCSs with fixed reset time instants. In this case, the reset controller is of the form

$$\begin{cases} \dot{x}_c = A_c x_c + B_c e, & t \in ((k-1)T_s, kT_s) \\ x_c^+ = \rho_c(x_c, r), & t = kT_s \\ u = C_c x_c + D_c e. \end{cases} \tag{1.45}$$

This reset controller can be viewed as a special case of (1.41) where ϕ is given in (1.42) with S a positive definite matrix.

The benefits of introducing discrete-time triggering conditions are multi-fold. First of all, Zeno phenomenon is naturally avoided and for any initial condition, a solution always exists. Second, discrete-time triggering conditions provide more flexibility in control design as explained in Chapters 5 and 6.

1.3 Fundamental theory of traditional reset design

1.3.1 Horowitz's design

Consider a RCS depicted in Figure 1.15, where $P(s)$ is the plant and the controller consists of a linear compensator $C(s)$ cascaded with a FORE. The baseline transfer function of the FORE is $\frac{1}{s+b}$, thus the FORE can be represented in state space as

$$\begin{cases} \dot{z} = -bz + e, & e \neq 0 \\ z^+ = 0, & e = 0 \\ u_r = z. \end{cases} \tag{1.46}$$

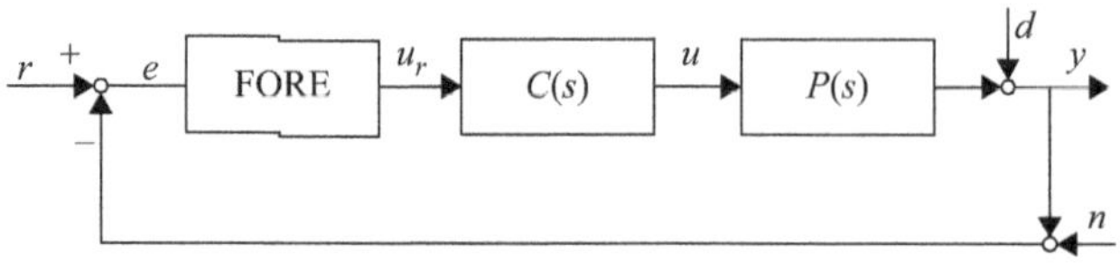

Figure 1.15 A FORE RCS

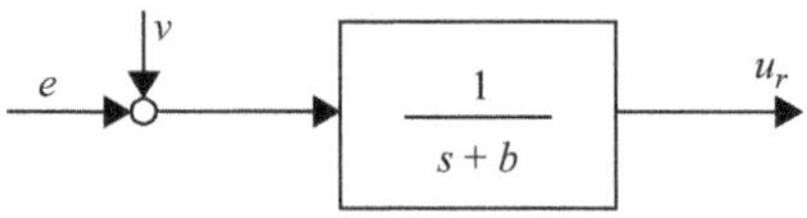

Figure 1.16 A FORE equivalence

The pole $-b$ of the FORE will be an adjustable parameter in the design procedure.

Denote by t_i the ith reset time instant with $0 = t_0 < t_1 < t_2 < \cdots < t_n \leq \infty$. n might be finite or infinite. In the case when $n = \infty$, there holds $t_n = \infty$. Suppose that $e(t) = 0$ for all $t < 0$. Similar to the discussion regarding the CI in Section 1.1, the output response of the FORE can be calculated as

$$\begin{aligned} u_r(t) &= e^{-bt} \int_{t_i}^{t} e^{bs} e(s) ds, \quad t \in [t_i, t_{i+1}) \\ &= e^{-bt} \left[\int_0^t e^{bs} e(s) ds - \sum_{t_k \leq t} \int_{t_{k-1}}^{t_k} e^{bs} e(s) ds \right] \\ &= e^{-bt} \int_0^t e^{bs} \left[e(s) + v(s) \right] ds \end{aligned}$$

where

$$\begin{aligned} v(t) &= -\sum_{k=1}^{n} A_k \delta(t - t_k) \\ A_k &= \int_{t_{k-1}}^{t_k} e^{-b(t_k - s)} e(s) ds. \end{aligned}$$

This means that the FORE can be alternatively represented as a linear system with an additional input $v(t)$ as depicted in Figure 1.16. Denote by $h_L(t)$ and $y_L(t)$, respectively, the impulse and the step responses of the baseline system, i.e., with the FORE being replaced by $\frac{1}{s+b}$, and by $y(t)$ the step response of the RCS (with $d = 0$ and $n = 0$). Then there holds

$$\begin{aligned} y(t) &= y_L(t) - \sum_{k=1}^{n} A_k h_L(t - t_k) \\ &= y_L(t) - \sum_{k=1}^{n} A_k \dot{y}_L(t - t_k). \end{aligned}$$

For $t \in [0, t_1]$, we have $y(t) = y_L(t)$ and $y_L(t_1) = 0$. For $t \in [t_1, t_2)$,

$$y(t) = y_L(t) - A_1 \dot{y}_L(t - t_1)$$

with

$$A_1 = \int_0^{t_1} e^{-b(t_1 - s)} [1 - y_L(s)] \, ds > 0.$$

From the above formula, if the step response of the baseline system is as depicted in Figure 1.9, there holds

$$\dot{y}_L(t) > 0, \quad t \in [0, \tilde{t}_1],$$

where $\tilde{t}_1$ represents the time instant of the first peak of $y_L(t)$. Thus we have

$$y(t) < y_L(t), \quad t \in [t_1, t_1 + \tilde{t}_1].$$

This intuitively explains that reset tends to reduce the overshoot.

If $b = 0$, then the FORE degenerates to the CI in which case A_1 is the area under the curve $e(t)$ within the interval $[0, t_1]$. In general, there holds $b \neq 0$. Thus, A_1 is the area under the curve $e^{-b(t_1 - t)} e(t)$ and the FORE pole plays the role of decreasing A_1. If the pole of the FORE is chosen to be $b \gg 0$, we have $A_k \approx 0$ and hence

$$y(t) \approx y_L(t), \quad t \geq 0.$$

This explains that the replacement of the CI by the FORE provides more flexibility in the control design.

If the baseline open-loop transfer function is the second-order system

$$L(s) = \frac{1}{s + b} P(s) C(s) = \frac{\omega_n^2}{s(s + 2\zeta \omega_n)}, \quad (0 < \zeta < 1), \tag{1.47}$$

then the overshoot of the RCSs is given by

$$\text{Overshoot} = \exp\left(\frac{-\pi \zeta}{\sqrt{1 - \zeta^2}} \right) - \Delta, \tag{1.48}$$

where

$$\Delta = \begin{cases} \dfrac{R}{1 - 4\zeta^2 M + 4\zeta^2 M^2} \left[4\zeta^2 M^2 e^{-\zeta\mu} - 2\zeta M (1 - 4\zeta^2 M) e^{-\frac{\mu}{2\zeta M}} \right], & \zeta \geq 0.5 \\ \dfrac{R}{1 - 2\zeta M + M^2} \left[M^2 e^{-\zeta\mu} - M(1 - 2\zeta M) e^{-\frac{\mu}{M}} \right], & \zeta \leq 0.5 \end{cases} \tag{1.49}$$

with

$$R = \exp\left(-\frac{\zeta}{\sqrt{1 - \zeta^2}} \arccos \zeta \right)$$

$$M = \frac{\omega_c}{b}, \quad \mu = \frac{\pi - \arccos \zeta}{\sqrt{1 - \zeta^2}}.$$

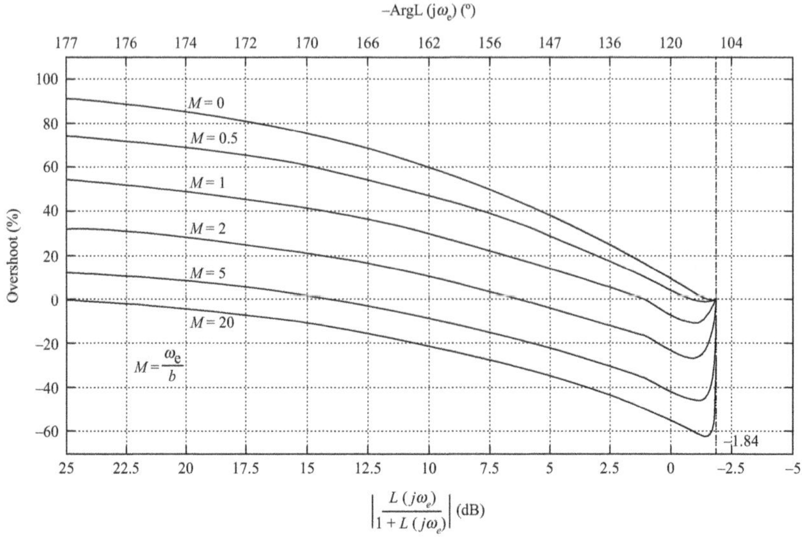

Figure 1.17 Overshoot with $M = \omega_c/b$ as parameter

In the above, ω_c is the gain-crossover frequency of the open-loop transfer function, i.e., $|L(j\omega_c)| = 1$. Denote by $G_L(s)$ the closed-loop transfer function, i.e.,

$$G_L(s) = \frac{L(s)}{1 + L(s)} = \frac{\omega_n^2}{s^2 + 2\zeta\omega_n s + \omega_n^2}$$

and define $D = |G_L(j\omega_c)|$. The damping ratio ζ can be related to D with

$$\zeta = \frac{1}{2D}\sqrt{\frac{4D^2 - 1}{2(D^2 - 1)}}. \tag{1.50}$$

Thus the overshoot of the second-order RCS can be represented as a function of D and M, which is plotted in Figure 1.17 with M the parameter. From this figure, when b is too small, undershoot occurs which explains the disadvantage of the CI.

Figure 1.17 is obtained from second-order systems but is also a good approximation for higher-order systems which are dominant by a second-order dynamics. This is the main tool in the Horowitz's design procedure.

The Horowitz's design [3] is depicted in Figure 1.18 where the reset controller consists of two parts, i.e., the linear compensator $C(s)$ and the FORE with pole $-b$. The linear controller $C(s)$ is of the form

$$C(s) = (s + b)C_L(s). \tag{1.51}$$

The above structure assures that the baseline dynamics does not change while adjusting the parameter b, i.e.,

$$L(s) = \frac{1}{s + b}P(s)C(s) = P(s)C_L(s).$$

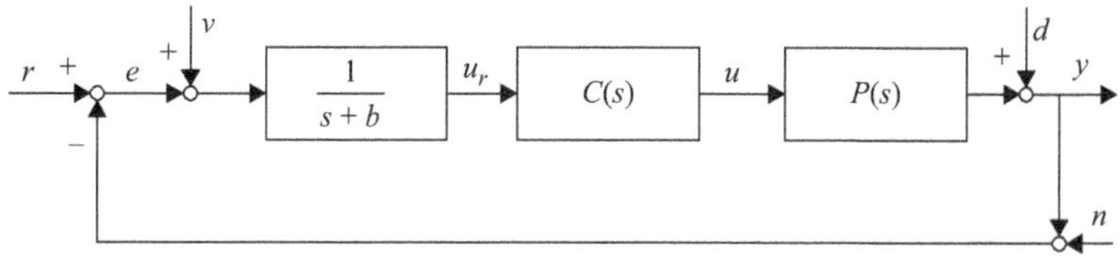

Figure 1.18 A FORE RCS equivalence

The Horowitz's design includes two steps:

- **Step 1:** Design a linear controller $C_L(s)$ such that the baseline system satisfies both the disturbance rejection and sensor noise suppression at the expense of violating the gain/phase margin constraint;
- **Step 2:** Choose a FORE pole $-b$ to reduce the overshoot.

The above design procedure was applied to a tape-speed control system in [14] where the plant is given by

$$P(s) = \frac{5.32 \times 10^5(s^2 - 6.95 \times 10^2 s + 2.16 \times 10^5)}{(s + 14.02)(s^2 + 7.451 \times 10^1 s + 2 \times 10^4)(s^2 + 2.03 \times 10^2 s + 3.28 \times 10^4)}.$$

The specifications for this system are as follows:

1. The disturbance attenuation level from d to e within the frequency range $\omega \leq 2$ Hz is not greater than 0.5, i.e.,

$$\left| \frac{1}{1 + L(j\omega)} \right| \leq 0.5, \quad \omega \leq 2\,\text{Hz}. \tag{1.52}$$

2. The sensor-noise suppression ratio for $\omega \geq 10$ Hz is not greater than 0.4, i.e.,

$$\left| \frac{L(j\omega)}{1 + L(j\omega)} \right| \leq 0.4, \quad \omega \geq 10\,\text{Hz}. \tag{1.53}$$

3. The steady-state error for step input is zero and the overshoot is not greater than 20%.

It has been shown that within the framework of linear feedback design, it is difficult to find a controller which meets all of the above specifications. By using the Horowitz's design, a linear controller $C_L(s)$ is first designed to meet the first two specifications with

$$\begin{aligned} C_L(s) &= \frac{11.3(s + 257.1)(s + 3.606)(s^2 + 20.33s + 216.3)}{s(s + 220.3)(s + 36.49)(s^2 + 14.29s + 62.48)} \\ &\quad \times \frac{(s^2 + 55.51s + 3679)(s^2 + 75.74s + 1.579 \times 10^4)}{(s^2 + 88.25s + 5925)(s^2 + 125.7s + 1.579 \times 10^4)}. \end{aligned}$$

The crossover frequency of $L(s) = P(s)C_L(s)$ is approximately $\omega_c \approx 30$ rad/s and $-\arg L(j\omega_c) \approx 150°$. From Figure 1.17, choosing a FORE pole $b = 30$ (i.e., $M = 1$)

reduces the overshoot to 20%. Thus the resulting reset controller consists of a FORE with $b = 30$ and a linear compensator

$$C(s) = (s + 30)G_L(s). \tag{1.54}$$

1.3.2 PI + CI reset design

Proportional-plus-integral control with Clegg integrator (PI + CI) reset control design is a DF-based design method aiming to improve the closed-loop response under a linear PI controller [19]. The main idea is to reset a percentage of the LI to retain the benefit of LI such as the capability of eliminating steady-errors, and at the same time increase the phase margin and the gain-crossing frequency.

The first step of the PI + CI reset controller design is to design a linear PI of the form

$$C(s) = K_p\left(1 + \frac{1}{T_I s}\right) \tag{1.55}$$

based on the desired phase margin and crossover frequency, where K_p is the proportional constant and T_I is the integral time constant. Then decompose the integral part as

$$\frac{1}{T_I s} = \frac{1}{T_i s} + \frac{1}{T_{ir} s}$$

and replace the second integrator in the above decomposition by a CI as depicted in Figure 1.19. The integral constants satisfy

$$\frac{1}{T_I} = \frac{1}{T_i} + \frac{1}{T_{ir}}$$

and the ratio T_i/T_{ir} characterizes the percentage of the integral to be reset. The DF of the PI + CI compensator is given by

$$C_r(j\omega) = K_p\left[1 + \frac{1}{j\omega T_i} + \frac{1}{j\omega T_{ir}}\left(1 + j\frac{4}{\pi}\right)\right]. \tag{1.56}$$

The remaining step is to tune the ratio T_i/T_{ir} according to the final performance requirement of the RCS.

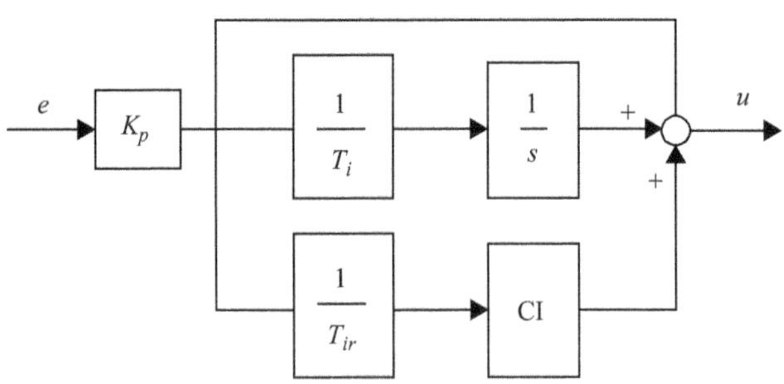

Figure 1.19 PI + CI reset controller

In the case when the plant is second order given by

$$P(s) = \frac{\omega_n^2}{(s^2 + \xi\omega_n + j\omega_d)(s^2 + \xi\omega_n - j\omega_d)}, \tag{1.57}$$

where ξ $(0 < \xi < 1)$ is the damping ratio, ω_n the undamped natural frequency and $\omega_d = \omega_n\sqrt{1-\xi^2}$, then the parameters of the PI + CI controller can be determined as follows.

1. **Linear PI tuning**. Based on the desired gain-crossover frequency ω_g and the phase margin ϕ_m, calculate T_I and K_p by

$$\begin{cases} \angle P(j\omega_g)C(j\omega_g) = \phi_m \\ |P(j\omega_g)C(j\omega_g)| = 1. \end{cases} \tag{1.58}$$

Or, precisely,

$$\begin{cases} \phi_m = \frac{1}{2}\pi + \arctan\omega_g T_I - \arctan\frac{\omega_g + \omega_d}{\xi\omega_n} - \arctan\frac{\omega_g - \omega_d}{\xi\omega_n} \\ \frac{K_p\omega_n^2}{\omega_g T_I}\sqrt{\frac{1+\omega_g^2 T_I^2}{[\xi^2\omega_n^2 + (\omega_g+\omega_d)^2][\xi^2\omega_n^2 + (\omega_g - \omega_d)^2]}} = 1 \end{cases} \tag{1.59}$$

2. **PI + CI parameter design**. Select a feasible crossover frequency ω_{gr} for the RCS and calculate T_{ir} according to

$$\begin{cases} |P(j\omega_{gr})C_r(j\omega_{gr})| = 1 \\ \frac{1}{T_I} = \frac{1}{T_i} + \frac{1}{T_{ir}}. \end{cases} \tag{1.60}$$

Simple calculation gives

$$T_{ir} = \frac{4}{\pi\left[\sqrt{\beta(\omega_{gr}) - 1/T_I^2} - \omega_{gr}\right]}, \tag{1.61}$$

where β is given by

$$\beta(\omega_{gr}) = \frac{\omega_{gr}^2}{K_p^2\omega_n^4}\left[\xi^2\omega_n^2 + (\omega_{gr}+\omega_d)^2\right]\left[\xi^2\omega_n^2 + (\omega_{gr}-\omega_d)^2\right]. \tag{1.62}$$

(1.61) gives the range of ω_{gr} one can choose from, i.e.,

$$\sqrt{\beta(\omega_{gr}) - 1/T_I^2} > \omega_{gr}.$$

The phase margin ϕ_{rm} of the PI + CI RCS is given by

$$\begin{aligned} \phi_{rm} &= \angle P(j\omega_{gr})C_r(j\omega_{gr}) \\ &= \frac{1}{2}\pi + \arctan\left[\omega_{gr} + 4/(T_{ir}\pi)\right]T_I - \arctan\frac{\omega_g + \omega_d}{\xi\omega_n} - \arctan\frac{\omega_g - \omega_d}{\xi\omega_n}. \end{aligned}$$

Notes

A detailed description of the limitations of linear feedback control can be found in Reference 1 and the example showing that reset control achieves a set of specifications which cannot be achieved by linear control comes from Reference 18. The early developments of reset control including the CI, the first time-domain design around the CI, and the FORE (the Horiwitz's design procedure) studied in this chapter are, respectively, from References 11, 16, and 3. The application example of Horiwitz's design procedure to the tape-speed control systems can be found in Reference 14. See also the References 13, 17, and 15 and the recently published book [20] for more detailed discussions of traditional reset control analysis. The basic PI + CI reset control design comes from Reference 19. See also Reference 21 for the application of PI + CI reset control design to heat exchangers.

The model (1.5) for the CI is from Reference 12. The definitions of solutions and the well-posedness of RCSs are based on Reference 20. For the theory of general hybrid systems and impulsive differential equations, reader can refer to References 22, 23, and 10. Other nonlinear techniques aiming to overcome limitations of linear feedback control can be found in References 24, 8, 9, 7, 4, and 6.

References

[1] M.M. Seron, J.H. Braslavsky, and G.C. Goodwin. *Fundamental limitations in filtering and control*. In *Communications and Control Engineering*. Springer-Verlag London Ltd., Berlin, Heidelberg, New York, 1997.

[2] J.S. Freudenberg and D.P. Looze. *Frequency domain properties of scalar and multivariable feedback systems*. Springer-Verlag New York, Inc., 1988.

[3] I. Horowitz and P. Rosenbaum. Nonlinear design for cost of feedback reduction in systems with large parameter uncertainty. *International Journal of Control*, 21(6):977–1001, 1975.

[4] J.B. Lewis. The use of nonlinear feedback to improve the transient response of a servomechanism. *American Institute of Electrical Engineers, Part II: Transactions of the Applications and Industry*, 71(6):449–453, 1953.

[5] E.R. Kalman. Phase-plane analysis of automatic control systems with nonlinear gain elements. *American Institute of Electrical Engineers, Part II: Applications and Industry, Transactions of the*, 73(6):383–390, 1955.

[6] A.R. Bailey. Stabilisation of control systems by the use of driven limiters. *Electrical Engineers, Proceedings of the Institution of*, 113(1):169–174, 1966.

[7] W.C. Foster, D.L. Gieseking, and W.K. Waymeyer. A nonlinear filter for independent gain and phase (with applications). *Journal of Basic Engineering*, 88:457, 1966.

[8] D. Liberzon. *Switching in systems and control*. In *Systems & Control: Foundations & Applications*, Birkhäuser, Boston, Basel, Berlin, 2003.

[9] C. Edwards and S. Spurgeon. *Sliding mode control: theory and applications*. In *Systems and Control*, CRC Press, Taylor & Francis Group, 1998.

[10] T. Yang. *Impulsive control theory*. In *Lecture Notes in Control and Information Sciences*, vol. 272, Springer-Verlag Berlin, Heidelberg, Springer, 2001.

[11] J.C. Clegg. A nonlinear integrator for servomechanisms. *Institute of Electrical Engineers, Part II: Transactions of the Applications and Industry*, 77(1):41–42, 1958.

[12] L. Zaccarian, D. Nesic, and A.R. Teel. First order reset elements and the Clegg integrator revisited. In *American Control Conference, 2005. Proceedings of the 2005*. June 8–10, 2005, vol. 1, pp. 563–568. IEEE, Portland, OR, 2005.

[13] O. Beker. *Analysis of reset control systems*. Doctoral dissertations, University of Massachusetts Amherst, *Available from Proquest*. Paper AAI3027178. http://scholarworks.umass.edu/dissertations/AAI3027178, January 1, 2001.

[14] Y. Zheng, Y. Chait, C.V. Hollot, M. Steinbuch, and M. Norg. Experimental demonstration of reset control design. *Control Engineering Practice*, 8(2):113–120, 2000.

[15] Y. Chait and C.V. Hollot. On Horowitz's contributions to reset control. *International Journal of Robust and Nonlinear Control*, 12(4):335–355, 2002.

[16] K.R. Krishnan and I.M. Horowitz. Synthesis of a nonlinear feedback system with significant plant-ignorance for prescribed system tolerances. *International Journal of Control*, 19(4):689–706, 1974.

[17] Q. Chen. *Reset control systems: stability, performance and application*. PhD thesis, University of Massachusetts Amherst, *Available from Proquest*. Paper AAI9988771. http://scholarworks.umass.edu/dissertations/AAI9988771, Januray 1, 2000.

[18] O. Beker, C.V. Hollot, and Y. Chait. Plant with integrator: an example of reset control overcoming limitations of linear feedback. *IEEE Transactions on Automatic Control*, 46(11):1797–1799, 2001.

[19] A. Baños and A. Vidal. Design of PI+CI reset compensators for second order plants. In *IEEE International Symposium on Industrial Electronics*, pp. 118–123. IEEE, 2007.

[20] A. Baños and A. Barreiro. *Reset control systems*. Springer Science & Business Media, 2011.

[21] A. Vidal and A. Baños. Reset compensation for temperature control: Experimental application on heat exchangers. *Chemical Engineering Journal*, 159(1–3):170–181, 2010.

[22] R. Goebel, R.G. Sanfelice, and A.R. Teel. *Hybrid dynamical systems: modeling, stability, and robustness*. Princeton University Press, 2012.

[23] W.M. Haddad, V. Chellaboina, and S.G. Nersesov. *Impulsive and hybrid dynamical systems: stability, dissipativity, and control*. In *Princeton Series in Applied Mathematics*, Princeton University Press, Princeton, New Jersey, 2006.

[24] Z. Sun and S.S. Ge. *Switched linear systems: control and design*. Springer-Verlag, Berlin, Heidelberg, 2005.

Chapter 2
Describing function analysis of reset systems

Describing function (DF) is a quasi-linearization of a nonlinear element subject to certain excitation input used to approximately analyze the behavior of nonlinear systems. The *sinusoidal DF*, which uses sinusoidal inputs as excitation signals, is most widely known. DF is a powerful tool in investigating behaviors of elements with hard nonlinearities including dead zone, backlash, and hysteresis, and has been applied in limit cycle prediction and control design. The basic motivation to study DFs of reset systems is that by far rigorous mathematical tools are not available to analyze the behavior of state-driven reset. This chapter first derives the sinusoidal DF for a general reset system with arbitrary reset matrix, followed by application of the DF in disturbance rejection of *hard disk drive* (HDD) systems.

2.1 Sinusoid input response

Consider the general reset system

$$\Sigma_r : \quad \begin{cases} \dot{x} = Ax + be, & e \neq 0 \\ x(t^+) = Rx, & e = 0 \\ y = c^T x, \end{cases} \tag{2.1}$$

where $e, y \in \mathbb{R}^1$ are the scalar input and output, respectively, x, A, b, R, c are of compatible dimensions. The second equation of (2.1) is called the *reset mapping* and R is called the *reset matrix*. $x(t^+)$ denotes the after-reset state at reset time instant t, i.e.,

$$x(t^+) \triangleq \lim_{s \to t+0} x(s).$$

System (2.1) can be a reset element or the open loop of a *reset control system* (RCS). For instance, consider an RCS depicted in Figure 2.1. The plant Σ_p and the reset controller Σ_{rc} are, respectively,

$$\Sigma_p : \quad \begin{cases} \dot{x}_p = A_p x_p + B_p u \\ y = c_p^T x_p \end{cases} \tag{2.2}$$

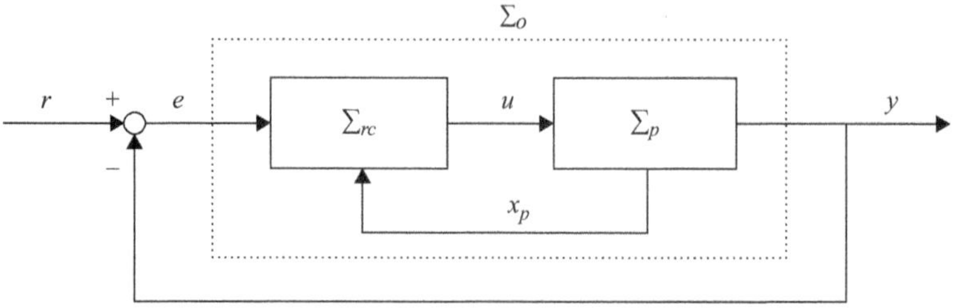

Figure 2.1 General RCS

and

$$\Sigma_{rc}: \begin{cases} \dot{z} = A_r z + A_{rp}x_p + B_r e, & e \neq 0 \\ z(t^+) = R_r z + R_{rp}x_p, & e = 0 \\ u = E_r z + E_{rp}x_p + E_e e. \end{cases} \tag{2.3}$$

Combining Σ_p and Σ_{rc}, the open-loop system Σ_o can be represented in the form of system (2.1) with $x = (x_p^T, z^T)^T$ and

$$A = \begin{bmatrix} A_p + B_p E_{rp} & B_p E_r \\ A_{rp} & A_r \end{bmatrix}, \quad B = \begin{bmatrix} B_p E_e \\ B_r \end{bmatrix}, \tag{2.4}$$

$$R = \begin{bmatrix} I & O \\ R_{rp} & R_r \end{bmatrix}, \quad c^T = \begin{bmatrix} c_p^T & O \end{bmatrix}. \tag{2.5}$$

Remark 2.1. *In the early development of reset systems, the reset matrix R is assumed to be zero, i.e., $R = 0$. See Reference 1 for instance. In this case, all of the states are reset to zero whenever the input crosses zero. From (2.5), the reset matrix R of the open-loop system is never zero, this is because only the state of the controller is allowed to be reset. In this chapter, we assume the reset matrix to be an arbitrary square matrix.*

For any matrices A and R, we introduce the following notations (if well-defined) for convenience

$$\Lambda(\omega) \triangleq \omega^2 I + A^2,$$

$$\Delta(\omega) \triangleq I + e^{\frac{\pi}{\omega}A},$$

$$\Delta_R(\omega) \triangleq I + Re^{\frac{\pi}{\omega}A},$$

$$\Gamma_R(\omega) = \Delta_R^{-1}(\omega) R \Delta(\omega) \Lambda^{-1}(\omega),$$

$$\Omega = \{\omega > 0 \mid \pm j\omega \text{ are eigenvalues of } A\},$$

$$\Omega_R = \{\omega > 0 \mid \left|\lambda(Re^{\frac{\pi}{\omega}A})\right| < 1\}.$$

The following assertions are obviously true:

1. $\Lambda(\omega)$ $(\omega > 0)$ is nonsingular if and only if $\omega \notin \Omega$.
2. If $\omega \in \Omega_R$, then $\Delta_R(\omega)$ is nonsingular.

In order to obtain the DF of system (2.1), we study a sinusoid input response first. Consider the sinusoid excitation input

$$e(t) = \alpha \sin(\omega t). \tag{2.6}$$

The set of the reset time instants $\{t_k\}$ is given by

$$t_k = k\pi/\omega, \quad k = 0, 1, \ldots.$$

Define

$$\begin{cases} \eta_k = x(t_{2k}^+), \\ \zeta_k = x(t_{2k+1}^+) \end{cases}$$

and

$$\psi(t) \triangleq \int_0^t e^{-As} b \sin(\omega s) ds.$$

Then the solution of system (2.1) with initial condition $x(0^+) = \eta_0$ can be represented by

$$x(t) = \begin{cases} e^{A(t-t_{2k})}\eta_k + \alpha e^{At}[\psi(t) - \psi(t_{2k})], & t \in (t_{2k}, t_{2k+1}]; \\ e^{A(t-t_{2k+1})}\zeta_k + \alpha e^{At}[\psi(t) - \psi(t_{2k+1})], & t \in (t_{2k+1}, t_{2k+2}] \end{cases} \tag{2.7}$$

where η_k and ζ_k are determined by the following recursive algorithm

$$\begin{cases} \zeta_k = Re^{\frac{\pi}{\omega}A}\left[\eta_k + \alpha\psi(\frac{\pi}{\omega})\right], \\ \eta_{k+1} = Re^{\frac{\pi}{\omega}A}\left[\zeta_k - \alpha\psi(\frac{\pi}{\omega})\right], \quad \eta_0 = x(0^+). \end{cases} \tag{2.8}$$

Definition 2.1. *The recursion (2.8) is said to converge globally if there exist constant vectors $\bar{\zeta}$ and $\bar{\eta}$ such that*

$$\lim_{k\to+\infty} \eta_k = \bar{\eta}, \tag{2.9}$$

$$\lim_{k\to+\infty} \zeta_k = \bar{\zeta} \tag{2.10}$$

for an arbitrary initial condition η_0.

The following lemma gives the necessary and sufficient condition for the global convergence of the recursion (2.8).

Lemma 2.1. *The recursion (2.8) converges globally if and only if $\omega \in \Omega_R$, or equivalently,*

$$\left|\lambda(Re^{\frac{\pi}{\omega}A})\right| < 1. \tag{2.11}$$

In this case, we have

$$\bar{\eta} \triangleq \lim_{k\to+\infty} \eta_k = -\alpha\left[I + Re^{\frac{\pi}{\omega}A}\right]^{-1} Re^{\frac{\pi}{\omega}A}\psi(\frac{\pi}{\omega}), \tag{2.12}$$

$$\bar{\zeta} \triangleq \lim_{k\to+\infty} \zeta_k = \alpha\left[I + Re^{\frac{\pi}{\omega}A}\right]^{-1} Re^{\frac{\pi}{\omega}A}\psi(\frac{\pi}{\omega}). \tag{2.13}$$

Proof: (Sufficiency) Assume that (2.11) holds, then $I + Re^{\frac{\pi}{\omega}A}$ is nonsingular. Let

$$\xi_k = \eta_k + \alpha \left[I + Re^{\frac{\pi}{\omega}A} \right]^{-1} Re^{\frac{\pi}{\omega}A} \psi(\frac{\pi}{\omega}),$$

we have

$$\xi_{k+1} = \left[Re^{\frac{\pi}{\omega}A} \right]^2 \xi_k. \tag{2.14}$$

Since

$$\left| \lambda \left[(Re^{\frac{\pi}{\omega}A})^2 \right] \right| = \left| \lambda(Re^{\frac{\pi}{\omega}A}) \right|^2 < 1,$$

we have

$$\lim_{k \to +\infty} \xi_k = 0.$$

Thus (2.12) holds. (2.13) can be proved similarly.

(Necessity) From (2.8) we have

$$\eta_{k+1} = \left[Re^{\frac{\pi}{\omega}A} \right]^2 \eta_k + \chi,$$

where

$$\chi = \alpha \left[Re^{\frac{\pi}{\omega}A} \right]^2 \psi(\frac{\pi}{\omega}) - \alpha Re^{\frac{\pi}{\omega}A} \psi(\frac{\pi}{\omega})$$

is a constant vector. Let $\bar{\eta} = \lim_{k \to \infty} \eta_k$, then we have

$$\left[Re^{\frac{\pi}{\omega}A} \right]^2 \bar{\eta} + \chi = \bar{\eta}.$$

Further, by defining

$$\xi_k = \eta_k - \bar{\eta},$$

we have

$$\lim_{t \to \infty} \xi_k = 0$$

for any ξ_0 and

$$\xi_{k+1} = \left[Re^{\frac{\pi}{\omega}A} \right]^2 \xi_k.$$

Thus condition (2.11) holds.

By Lemma 2.1, we have the following result immediately.

Proposition 2.1. *Reset system (2.1) with input (2.6) has a globally asymptotically stable $2\pi/\omega$-periodic solution if and only if*

$$\left| \lambda(Re^{\frac{\pi}{\omega}A}) \right| < 1. \tag{2.15}$$

If the above condition holds, then the globally asymptotically stable solution is given by

$$x(t) = \begin{cases} e^{A(t-t_{2k})}\bar{\eta} + \alpha e^{At}[\psi(t) - \psi(t_{2k})], & t \in (t_{2k}, t_{2k+1}], \\ e^{A(t-t_{2k+1})}\bar{\zeta} + \alpha e^{At}[\psi(t) - \psi(t_{2k+1})], & t \in (t_{2k+1}, t_{2k+2}] \end{cases}$$

where $\bar{\eta}$ and $\bar{\zeta}$ are defined in (2.12) and (2.13).

Remark 2.2. *From Proposition 2.1, the existence of globally asymptotically periodic solution of an RCS under a sinusoid input with certain frequency ω does not imply that the base linear system is stable. In addition, condition (2.15) depends on input frequency ω. But for a linear system without reset (i.e., $R = I$), condition (2.15) implies that A must be a stable matrix. In this case, condition (2.15) is equivalent to* $\text{Re}\,\lambda(A) < 0$ *which is independent of ω.*

Remark 2.3. *From equation (2.14), $|\lambda(Re^{\frac{\pi}{\omega}A})|$ is related to the convergence rate of the periodic solution. This can also be seen in the following example and the simulation result of Section 2.3.*

Example 2.1. *Consider a reset system with*

$$A = \begin{bmatrix} -1 & 1 \\ 0 & 1 \end{bmatrix}, \quad b = \begin{bmatrix} 0 \\ 1 \end{bmatrix}, \quad c = \begin{bmatrix} 0 & 1 \end{bmatrix}, \quad R = \begin{bmatrix} 1 & 0 \\ 0 & \beta \end{bmatrix}.$$

It is easy to check that

$$Re^{\frac{\pi}{\omega}A} = \begin{bmatrix} e^{-\frac{\pi}{\omega}} & \frac{1}{2}[e^{\frac{\pi}{\omega}} - e^{-\frac{\pi}{\omega}}] \\ 0 & \beta e^{\frac{\pi}{\omega}} \end{bmatrix}.$$

Thus for any fixed input frequency $\omega > 0$, the condition (2.15) holds if and only if

$$|\beta| < e^{-\frac{\pi}{\omega}}.$$

The output responses of this system for initial condition $(5, -1)$ *with* $\beta = -0.3$ *and* 0, *respectively, are given in Figure 2.2.*

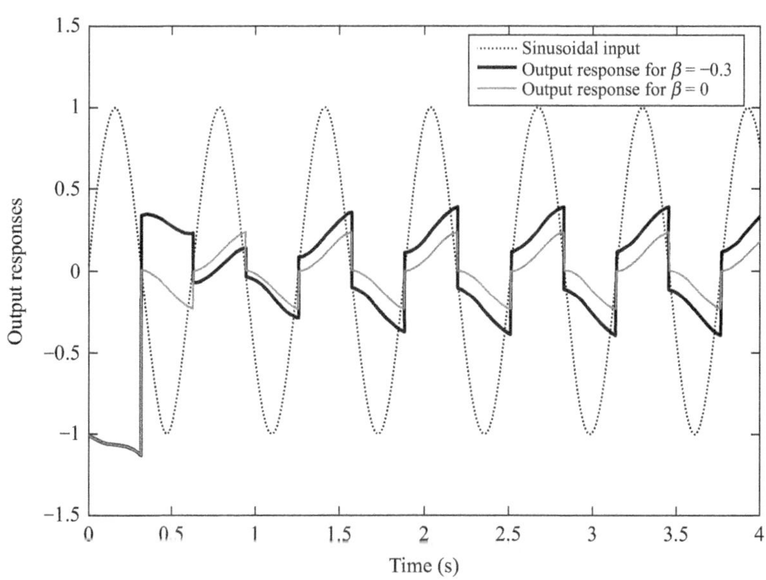

Figure 2.2 Output responses for Example 2.1

In practical systems, the actual input contains signals of different frequencies. Thus the boundedness of solutions under a sinusoid input with arbitrary frequency is an important property for RCSs. And this property is also the basis for a nonlinear element to have a DF defined on arbitrary frequency. From Proposition 2.1, the following result is obvious.

Proposition 2.2. *RCS (2.1) has a globally asymptotically stable $2\pi/\omega$-periodic solution under a sinusoid input with arbitrary frequency $\omega > 0$ if and only if*

$$|\lambda(Re^{A\delta})| < 1, \quad \forall \delta \in \mathbb{R}^+. \tag{2.16}$$

Remark 2.4. *From condition (2.16), by letting $\delta \to 0^+$, one sees that R must be such that $|\lambda(R)| \le 1$. However, A is not necessarily Lyapunov stable. For instance, in Example 2.1, A is unstable. But if $\beta = 0$, then*

$$Re^{A\delta} = \begin{bmatrix} e^{-A\delta} & \frac{1}{2}[e^{A\delta} - e^{-A\delta}] \\ 0 & 0 \end{bmatrix}$$

and condition (2.16) holds.

Remark 2.5. *If condition (2.16) holds, then system (2.1) has a sinusoid input DF defined on arbitrary frequency.*

Remark 2.6. *If $\omega \notin \Omega$, then $\omega^2 I + A^2$ is nonsingular. In this case, it is easy to check that*

$$\psi(t) = -[\omega^2 I + A^2]^{-1} \left[(\omega I \cos(\omega t) + A \sin(\omega t))e^{-At} - \omega I\right] b.$$

Thus

$$\psi\left(\frac{\pi}{\omega}\right) = [\omega^2 I + A^2]^{-1}(I + e^{-\frac{\pi}{\omega}A})\omega b. \tag{2.17}$$

2.2 Describing function

2.2.1 General case

In this section, we calculate the DF of system (2.1). According to the analysis of the previous section, the response $y(t)$ of system (2.1) under a sinusoid input with frequency $\omega \in \Omega_R$ can be represented as

$$y(t) = y_{ss}(t) + y_t(t), \tag{2.18}$$

where $y_{ss}(t)$ is the steady-state response which is $2\pi/\omega$-periodic and independent of the initial condition, and $y_t(t)$ is the transient response which dies away with time, i.e.,

$$\lim_{t\to\infty} y_t(t) = 0. \tag{2.19}$$

The *sinusoidal DF* of system (2.1) is defined by

$$G_R(j\omega) = \frac{Y_{ss,1}(j\omega)}{E(j\omega)}, \tag{2.20}$$

where

$$Y_{ss,1}(j\omega) = \frac{\omega}{2\pi}\int_0^{\frac{2\pi}{\omega}} y_{ss}(t)e^{-j\omega t}dt, \tag{2.21}$$

$$E(j\omega) = \frac{\omega}{2\pi}\int_0^{\frac{2\pi}{\omega}} e(t)e^{-j\omega t}dt. \tag{2.22}$$

In the following, we assume that $\omega \in \Omega_R \setminus \Omega$. By (2.7), (2.8) and (2.17), when $t \in (t_k, t_{k+1}]$,

$$\begin{aligned} y_{ss}(t) &= \alpha c^T e^{At}\left[(-1)^{k+1}e^{-At_k}\Gamma_R(\omega)\omega b + \psi(t) - \psi(t_k)\right] \\ &= \alpha c^T e^{At}\theta_k(\omega) - \alpha c^T\Lambda^{-1}(\omega)[\omega I\cos(\omega t) + A\sin(\omega t)]b \end{aligned}$$

where

$$\begin{aligned} \theta_k(\omega) &\triangleq (-1)^{k+1}e^{-At_k}\Gamma_R(\omega)\omega b - \psi(t_k) + \Lambda^{-1}(\omega)\omega b \\ &= (-1)^{k+1}e^{-At_k}[\Gamma_R(\omega) - \Lambda^{-1}(\omega)]\omega b. \end{aligned}$$

Thus the fundamental component of the Fourier series of $y_{ss}(t)$ can be calculated as follows

$$\begin{aligned} Y_{ss,1}(j\omega) &= \frac{\omega}{2\pi}\int_0^{\frac{2\pi}{\omega}} y_{ss}(t)e^{-j\omega t}dt \\ &= \frac{\alpha\omega c^T}{2\pi}(I_1 + I_2) - \frac{\alpha\omega c^T\Lambda^{-1}(\omega)}{2\pi}(\omega I J_1 + A J_2)b, \end{aligned}$$

where

$$\begin{aligned} I_1 &= \int_0^{\frac{\pi}{\omega}} e^{At}\theta_0(\omega)e^{-j\omega t}dt = (j\omega I - A)^{-1}\Delta(\omega)\theta_0(\omega), \\ I_2 &= \int_{\frac{\pi}{\omega}}^{\frac{2\pi}{\omega}} e^{At}\theta_1(\omega)e^{-j\omega t}dt \\ &= -(j\omega I - A)^{-1}\Delta(\omega)e^{\frac{\pi}{\omega}A}\theta_1(\omega), \\ J_1 &= \int_0^{\frac{2\pi}{\omega}} e^{-j\omega t}\cos(\omega t)dt = \frac{\pi}{\omega}, \\ J_2 &= \int_0^{\frac{2\pi}{\omega}} e^{-j\omega t}\sin(\omega t)dt = -j\frac{\pi}{\omega}. \end{aligned}$$

Thus we have

$$Y_{ss,1}(j\omega) = -\frac{j\alpha}{2}c^T(j\omega I - A)^{-1}(I + j\Theta_R(\omega))b \tag{2.23}$$

where

$$\Theta_R(\omega) \triangleq -\frac{2\omega^2}{\pi}\Delta(\omega)[\Gamma_R(\omega) - \Lambda^{-1}(\omega)] \tag{2.24}$$

$$= \frac{2\omega^2}{\pi}(I + e^{\frac{\pi}{\omega}A})(I + Re^{\frac{\pi}{\omega}A})^{-1}(I - R)(\omega^2 I + A^2)^{-1} \tag{2.25}$$

It is very easy to calculate that

$$E(j\omega) \triangleq \frac{\omega}{2\pi}\int_0^{\frac{2\pi}{\omega}} \alpha \sin(\omega t)e^{-j\omega t}dt = -\frac{j\alpha}{2}.$$

The following result follows.

Theorem 2.1. *The sinusoidal DF of system (2.1) is given by*

$$G_R(j\omega) = c^T(j\omega I - A)^{-1}(I + j\Theta_R(\omega))b \tag{2.26}$$

which is well-defined on $\omega \in \Omega_R \setminus \Omega$, where $\Theta_R(\omega)$ is given by (2.25).

Remark 2.7. *If the output function is replaced by*

$$y = c^T x + de, \tag{2.27}$$

then the DF becomes

$$G_R(j\omega) = c^T(j\omega I - A)^{-1}(I + j\Theta_R(\omega))b + d. \tag{2.28}$$

Remark 2.8. *If A has no pure imaginary eigenvalues, i.e., $\Omega = \emptyset$, and condition (2.16) holds, then we have $\Omega_R \setminus \Omega = \mathbb{R}^+$. That is to say, the DF is well-defined for an arbitrary positive number ω.*

Remark 2.9.

1. *When $R = I$, system (2.1) degenerates to the base linear system. In this case, by (2.25), it is easy to verify that $\Theta_R(\omega) = 0$. Hence the DF degenerates to the frequency response function of the base linear system.*
2. *If A is a stable matrix, we have*

   $$\lim_{\omega \to 0} e^{\frac{\pi}{\omega}A} = 0.$$

   Thus, $\lim_{\omega\to 0}\Theta_R(\omega) = 0$. This means that the reset does not have much effect over the very low-frequency range $\omega \ll \pi\|A\|$.
3. *It is easy to check that*

   $$\Theta_R(\infty) \triangleq \lim_{\omega\to\infty}\Theta_R(\omega) = \frac{4}{\pi}(I + R)^{-1}(I - R).$$

   This indicates that the effect of the reset is almost independent of frequency in a high-frequency region with $\omega \gg \pi\|A\|$.

Remark 2.10. *Though system (2.1) is essentially a nonlinear system, the DF is independent of the amplitude α of the excitation signal.*

Corollary 2.1. *If the reset matrix is zero, i.e., $R = 0_{n\times 0}$, then the sinusoidal DF of system (2.1) is given by*

$$G_R(j\omega) = c^T(j\omega I - A)^{-1}\left[I + j\frac{2\omega^2}{\pi}(I + e^{\frac{\pi}{\omega}A})(\omega^2 I + A^2)^{-1}\right]b \tag{2.29}$$

which is well-defined on $\omega \in \mathbb{R}^+ \setminus \Omega$. In addition, there holds

$$\lim_{\omega\to\infty} G_R(j\omega) = \left(1 + j\frac{4}{\pi}\right)G_L(j\omega) \tag{2.30}$$

where

$$G_L(j\omega) = c^T(j\omega I - A)^{-1}b$$

is the frequency response function of the base linear system.

By using Corollary 2.1, the DF of the *first-order reset element* (FORE)

$$\begin{cases} \dot{z} = -bz + e, & e \neq 0 \\ z^+ = 0, & e = 0 \\ u_r = z. \end{cases} \tag{2.31}$$

is given by

$$G_{FORE}(j\omega) = \frac{1}{j\omega + b}\left[1 + j\frac{2\omega^2(1 + e^{-\pi b/\omega})}{\pi(\omega^2 + b^2)}\right]. \tag{2.32}$$

Let $b = 0$, then we get the DF of the *Clegg integrator* (CI) as

$$G_{CI}(j\omega) = \frac{1}{j\omega}\left(1 + j\frac{4}{\pi}\right).$$

Example 2.2. *[2] Consider a second-order plant $y(s) = P(s)u(s)$ with*

$$P(s) = \frac{s+1}{s(s+0.2)}$$

controlled by a FORE

$$\begin{cases} \dot{x}_r = -x_r + e, & e \neq 0 \\ z^+ = 0, & e = 0 \\ u = x_r \end{cases} \tag{2.33}$$

where $e = r - y$. The open-loop dynamics is then of the form (2.1) with

$$A = \begin{bmatrix} -0.2 & 0 & 2 \\ 1 & 0 & 0 \\ 0 & 0 & -1 \end{bmatrix}, \quad b = \begin{bmatrix} 0 \\ 0 \\ 1 \end{bmatrix}, \quad c = [0.5 \;\; 0.5 \;\; 0], \quad R = \begin{bmatrix} 1 & 0 & 0 \\ 0 & 1 & 0 \\ 0 & 0 & 0 \end{bmatrix}.$$

The transfer function of the baseline open-loop system is then given by

$$G_L(s) = c^T(sI - A)^{-1}b = \frac{5}{s(5s+1)}.$$

By using Theorem 2.1, the DF of the open-loop RCS is

$$G_R(j\omega) = G_L(j\omega)\left[1 + j\frac{2\omega^2(1+e^{-\pi/\omega})}{\pi(\omega^2+1)}\right].$$

It is worth noting that this DF can also be directly obtained by

$$G_R(j\omega) = P(j\omega)G_{FORE}(j\omega) \tag{2.34}$$

where $G_{FORE}(j\omega)$ is given by (2.32) with $b = 1$. In this example, we have

$$\Theta_R(\omega) = \frac{2\omega^2(1+e^{-\pi/\omega})}{\pi(\omega^2+1)} > 0, \quad \forall\omega > 0.$$

This means that the reset action provides additional phase margin.

In the following, we consider the *power spectrum* of the steady-state output $y_{ss}(t)$ for a given input signal with certain frequency ω. Denote by $Y_{ss,n}(jn\omega)$ the nth harmonic component of the steady-state output $y_{ss}(t)$, i.e.,

$$Y_{ss,k}(jk\omega) = \frac{\omega}{2\pi}\int_0^{\frac{2\pi}{\omega}} y_{ss}(t)e^{-jk\omega t}dt, \quad k = 1, 2, \ldots.$$

Then the *power spectrum* is given by

$$p_k = |Y_{ss,k}(jk\omega)|^2, \quad k = 1, 2, \ldots.$$

By definition, we have

$$\begin{aligned} Y_{ss,k}(jk\omega) &= \frac{\omega}{2\pi}\int_0^{\frac{2\pi}{\omega}} y_{ss}(t)e^{-jk\omega t}dt \\ &= \frac{\alpha\omega c^T}{2\pi}(I_1 + I_2) - \frac{\alpha\omega c^T\Lambda^{-1}(\omega)}{2\pi}(\omega I J_1 + AJ_2)b, \end{aligned}$$

where

$$I_1 = \int_0^{\frac{\pi}{\omega}} e^{At}\theta_0(\omega)e^{-jk\omega t}dt = (jk\omega I - A)^{-1}\left[I - (-1)^k e^{\frac{\pi}{\omega}A}\right]\theta_0(\omega),$$

$$\begin{aligned} I_2 &= \int_{\frac{\pi}{\omega}}^{\frac{2\pi}{\omega}} e^{At}\theta_1(\omega)e^{-jk\omega t}dt \\ &= (-1)^k(jk\omega I - A)^{-1}\left[I - (-1)^k e^{\frac{\pi}{\omega}A}\right]e^{\frac{\pi}{\omega}A}\theta_1(\omega), \end{aligned}$$

$$J_1 = \int_0^{\frac{2\pi}{\omega}} e^{-jk\omega t}\cos(\omega t)dt = \begin{cases} \dfrac{\pi}{\omega}, & k = 1 \\ 0, & k \geq 2, \end{cases}$$

$$J_2 = \int_0^{\frac{2\pi}{\omega}} e^{-jk\omega t} \sin(\omega t) dt = \begin{cases} -j\dfrac{\pi}{\omega}, & k = 1 \\ 0, & k \geq 2. \end{cases}$$

Thus, for any $k \geq 2$, there holds

$$\begin{aligned} Y_{ss,k}(j\omega) &= \frac{-1+(-1)^k}{2} \frac{\alpha\omega^2}{\pi} c^T [jk\omega I - A]^{-1} \left[I - (-1)^k e^{\frac{\pi}{\omega}A} \right] [\Gamma_R(\omega) - \Lambda^{-1}(\omega)] b \\ &= \begin{cases} 0, & k = 2, 4, \ldots \\ \dfrac{\alpha}{2} c^T [jk\omega I - A]^{-1} \Theta_R(\omega) b, & k = 3, 5, \ldots. \end{cases} \end{aligned} \tag{2.35}$$

By using (2.23) and (2.35), the power spectrum can be easily calculated. Especially, for the first-order reset system

$$\begin{cases} \dot{x} = ax + be, & e \neq 0 \\ x^+ = Rx, & e = 0 \\ y = x \end{cases} \tag{2.36}$$

we have the following.

Theorem 2.2. *Suppose that the input signal to the reset system (2.36) is* $e = \alpha \sin \omega t$. *Then the power spectrum of the steady-state output is given by*

$$p_k = \begin{cases} \dfrac{\alpha^2 b^2}{4} \dfrac{1 + \Theta_R^2(\omega)}{\omega^2 + a^2}, & k = 1 \\ \dfrac{\alpha^2 b^2}{4} \dfrac{\Theta_R^2(\omega)}{k^2\omega^2 + a^2}, & k = 3, 5, \ldots \\ 0, & k = 2, 4, \ldots. \end{cases} \tag{2.37}$$

where

$$\Theta_R(\omega) = \frac{2}{\pi} \frac{1 + e^{a\pi/\omega}}{1 + (a/\omega)^2} \frac{1 - R}{1 + Re^{a\pi/\omega}}.$$

Define the spectrum of the power percentage as

$$\sigma_k(R, \omega) = \frac{|Y_{ss,k}(j\omega)|^2}{\sum_{n=1}^{\infty} |Y_{ss,n}(jn\omega)|^2} \times 100\%, \quad k = 1, 2, \ldots. \tag{2.38}$$

Thus $\sigma_1(R, \omega)$ characterizes the degree of approximation of the steady-state output by the fundamental component. In practice, we can use the approximate power percentage defined by

$$\sigma_k(R, \omega) = \frac{|Y_{ss,k}(j\omega)|^2}{\sum_{n=1}^{N} |Y_{ss,n}(jn\omega)|^2} \times 100\% \tag{2.39}$$

for a given positive integer N. The power ratio of the first-order reset system (2.36) with $a = -1$, $b = 1$ and $\omega = 1 \times 10^3$ rad for different reset matrix R is depicted in Figure 2.3. From this figure, one sees that the power ratio of the high-frequency harmonics increases while the reset coefficient $R \in [-1, 1]$ decreases.

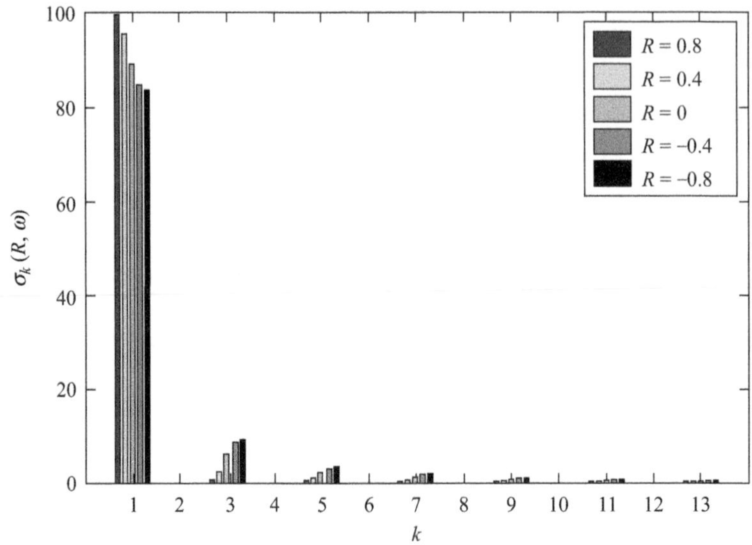

Figure 2.3 The spectrum of the power percentage $\sigma_k(R, \omega)$

2.2.2 Gain-balanced FORE

Consider the FORE

$$\begin{cases} \dot{z} = -2\pi f_r z + e, & e \neq 0 \\ z(t^+) = Rz, & e = 0 \\ y = \kappa(R)z \end{cases} \tag{2.40}$$

with $z, e, y \in \mathbb{R}^1$, $f_r > 0$. $\kappa(R) > 0$ is a scalar function of R used to balance the high-frequency gain caused by reset and will be determined later. $\kappa(R)$ is assumed to satisfy the condition $\kappa(1) = 1$. Thus when $R = 1$, (2.40) degenerates to an ordinary first-order system. We call the FORE (2.40) the *Gain-balanced FORE* (GFORE). According to the analysis in Sections 2.1 and 2.2, we have the following corollaries.

Corollary 2.2. *The system (2.40) with the sinusoid input $e = \alpha \sin(\omega t)$ has a globally asymptotically stable $2\pi/\omega$-periodic solution if and only if*

$$|R|e^{-2\pi^2 f_r/\omega} < 1.$$

Corollary 2.3. *The following two assertions are true:*

1. *The system (2.40) has a globally asymptotically stable $2\pi/\omega$-periodic solution under the sinusoid input $e = \alpha \sin(\omega t)$ with arbitrary frequency $\omega > 0$ if and only if $|R| \leq 1$.*
2. *The system (2.40) is bounded input bounded output stable (BIBOS) if and only if $|R| \leq 1$.*

Proof: The first claim follows directly from Proposition 2.2. The sufficiency of the second claim is obvious and we only need to prove the necessity. Suppose that the system (2.40) is BIBOS but $|R| > 1$. Thus we can always choose a frequency ω_0 large enough so that

$$|R|e^{-2\pi^2 f_r/\omega} > 1.$$

By Lemma 2.1, the recursion (2.8) diverges. Thus for the excitation input $e(t) = \sin \omega_0 t$, the output of the system (2.40) diverges which contradicts the BIBO stability.

According to Theorem 2.1, the DF of the GFORE (2.40) is

$$G_R(j\omega) = \frac{\kappa(R)\,[\,1 + j\Theta_R(\omega)\,]}{j\omega + 2\pi f_r},$$

where

$$\Theta_R(\omega) = \frac{2}{\pi}\frac{1 + e^{-2\pi^2 f_r/\omega}}{1 + (2\pi f_r/\omega)^2}\frac{1 - R}{1 + Re^{-2\pi^2 f_r/\omega}}.$$

By Corollaries 2.2 and 2.3, when $|R| > 1$, the DF $G_R(j\omega)$ is well-defined on the frequency range

$$0 < \omega < \frac{2\pi^2 f_r}{\ln |R|}.$$

When $|R| \leq 1$, it is defined for any arbitrary positive frequency $\omega > 0$.

The effect of reset on the frequency domain properties is characterized by $1 + j\Theta_R(j\omega)$. It is easy to verify that $\Theta_R(\omega) > 0$ if and only if

$$-e^{2\pi^2 f_r/\omega} < R < 1. \tag{2.41}$$

In other words, reset causes phase lead if and only if (2.41) holds. When $R = 1$, $\Theta_R = 0$, thus reset action does not cause phase lead nor phase lag. When $R > 1$ or $R < -e^{2\pi^2 f_r/\omega}$, we have $\Theta_R < 0$, thus reset action actually causes phase lag. In practice, we usually use the reset element to provide extra phase lead. Note that (2.41) holds for arbitrary frequency ω if and only if $|R| < 1$. Thus, in general, R is designed to satisfy $|R| < 1$. The gain and phase lead caused by resetting are, respectively, given by

$$g(R, \omega) \triangleq \kappa(R)|1 + j\Theta_R(\omega)|,$$

$$\varphi(R, \omega) \triangleq \arctan\,(\Theta_R(\omega)).$$

It is easy to show that

$$\lim_{\omega \to \infty} g(R, \omega) = \kappa(R)\sqrt{1 + \varrho^2(R)},$$

$$\lim_{\omega \to \infty} \varphi(R, \omega) = \arctan\,(\varrho(R))$$

where

$$\varrho(R) = \frac{4}{\pi} \cdot \frac{1 - R}{1 + R}. \tag{2.42}$$

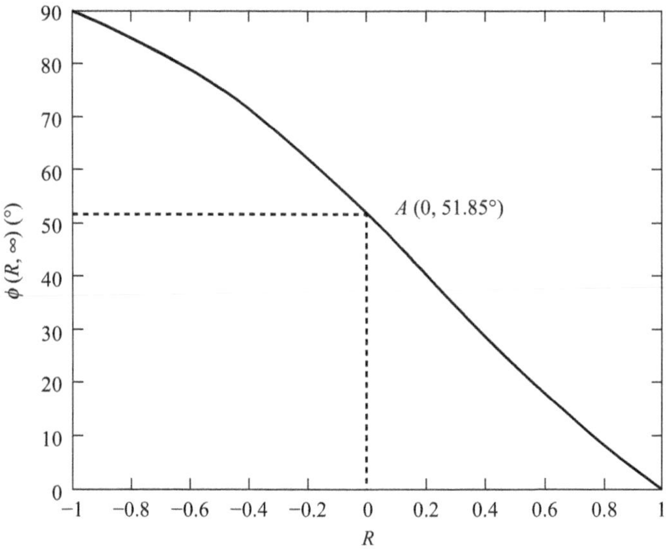

Figure 2.4 The phase lead of GFORE in large frequency range: $\varphi = \varphi(R, \infty)$

If we choose

$$\kappa(R) = \frac{1}{\sqrt{1 + \varrho^2(R)}} \tag{2.43}$$

to balance the gain caused by reset in high-frequency range, then in the high-frequency region $\omega \gg 2\pi f_r$, the gain properties of the overall GFORE for different R are almost the same, but the phase property is very sensitive to R. For example, when $R = 0$, (i.e., the traditional reset scheme), the phase lead caused by reset is about 51.9°. When $R = -0.9$, the phase lead caused by reset is about 87.6° (Figure 2.4 gives the relationship between R and the phase lead it generates in high-frequency range). This property allows us to change the phase property by choosing R without changing of the gain property in high frequency. Figure 2.5 gives the Bode plot of GFORE with $f_r = 0.1$. It shows that R only affects the phase property but does not change the gain property in a very large range of frequency.

From Figure 2.5, we can see that small R decreases the low-frequency gain. In this case, we can choose small f_r to compensate for low-frequency gain. See Figure 2.6. Actually, in the simulation of the next section, it can be seen that we can even choose $f_r = 0$ in which case the GFORE degenerates to the generalized CI and the gain is independent of R. See the following Remark 2.11.

Remark 2.11. *If $f_r = 0$, the GFORE becomes the Gain-balanced CI. In this case, the DF is*

$$G_R(j\omega) = \frac{\kappa(R)[1 + j\varrho(R)]}{j\omega}.$$

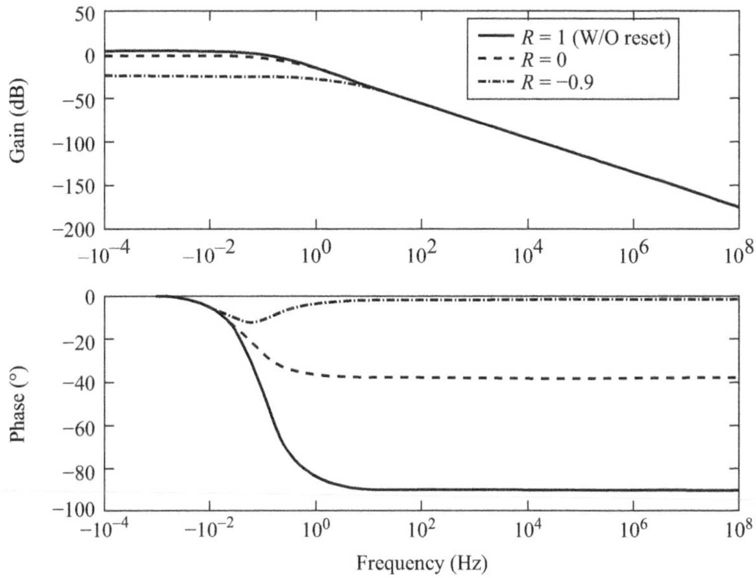

Figure 2.5 Bode plot of GFORE with $f_r = 0.1$ (based on DF)

Obviously, the gain

$$|G_R(j\omega)| = \frac{1}{|\omega|}$$

is independent of R, but the phase

$$\arg\left(G_R(j\omega)\right) = \arctan\left(\varrho(R)\right) - 90^o$$

is independent of frequency ω and only determined by R.

2.3 Application to HDD systems

2.3.1 Reset narrow band compensator (RNBC)

A *narrow band compensator* (NBC) has the form

$$G_{NBC}(s) = \frac{s^2 + 4\pi\xi_1 fs + 4\pi^2 f^2}{s^2 + 4\pi\xi_2 fs + 4\pi^2 f^2}, \quad \xi_1 > \xi_2, \tag{2.44}$$

where f is the frequency one wants to compensate for. The ratio ξ_1/ξ_2 determines the gain at f. An NBC can be inserted into the open-loop system to provide a large open-loop gain at certain disturbance frequency. However, direct insertion of an NBC

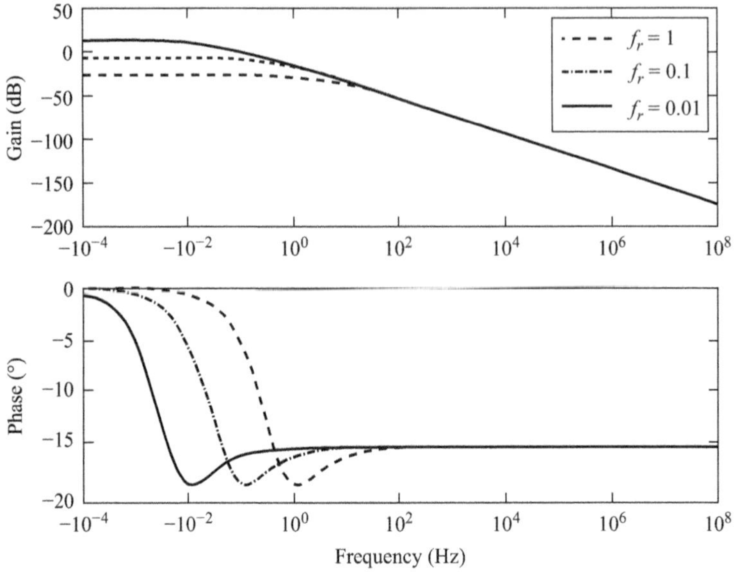

Figure 2.6 Bode plot of GFORE with R = −0.5 (based on DF)

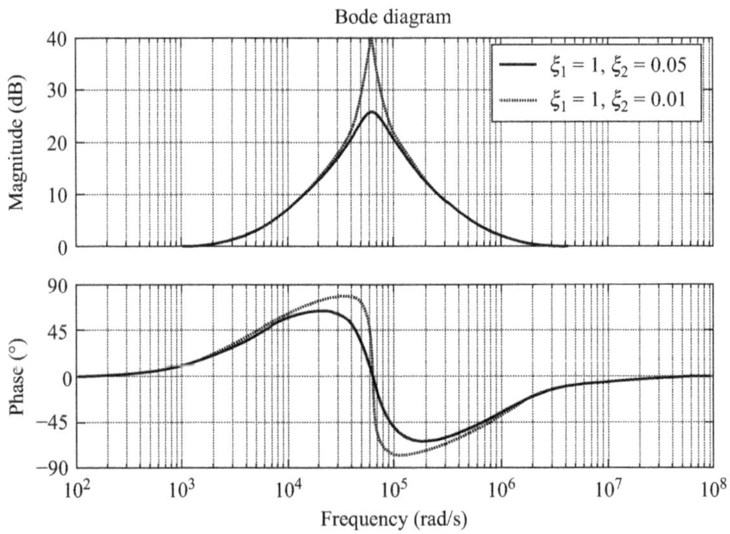

Figure 2.7 Bode diagram of the NBCs with $f = 1 \times 10^4$ and different ξ_1/ξ_1.

would result in phase margin loss because of the phase lag it causes. See Figure 2.7 for the Bode diagram of the NBC with

$$\xi_1 = 1, \quad \xi_2 = 0.05, \quad f = 1 \times 10^4.$$

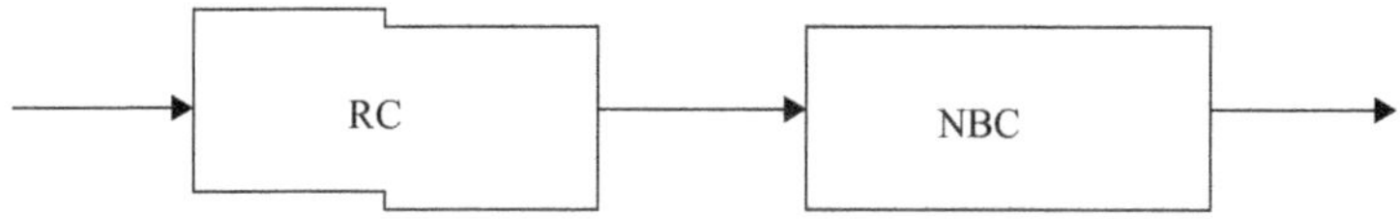

Figure 2.8 Reset narrow band compensator

The larger the ratio ξ_1/ξ_2, the larger the phase lag. It is easy to show that the largest phase lag caused by an NBC occurs at

$$\omega_0 = 2\pi f(\sqrt{\xi_1\xi_2} + \sqrt{\xi_1\xi_2 + 1}) \quad (\text{rad/s}) \tag{2.45}$$

and the largest phase lag is less than 90°.

The main idea of the *reset narrow band compensator* (RNBC) is to connect an NBC with a reset phase lead controller which is depicted in Figure 2.8. The reset controller (RC) is given by

$$\begin{cases} \dot{z} = -2\pi f_r z + e, & e \neq 0, \\ z(t^+) = Rz, & e = 0, \\ y_z = \kappa(R)(k_p\dot{z} + k_i z), & k_i \ll k_p. \end{cases} \tag{2.46}$$

The base linear system of the above reset system is in fact an ordinary linear phase lead mode. The reset action would provide an extra phase lead without changing the high-frequency magnitude.

2.3.2 Mid-frequency disturbance compensation

HDD is an important data storage device for modern computers. Figure 2.9 gives the structure of a typical commercial HDD servo system. The disk is coated with magnetic material used to store data which are arranged as concentric tracks. The *read/write* (R/W) head is mounted at the top of the arm which is actuated by a *voice coil motor* (VCM). In the working status, the disk rotates with high speed while the VCM provides the R/W head radial motion so that the R/W head can visit every data section on the disk.

Track-seeking and *track-following* are two basic functions of HDD servo systems. The task of track-seeking is to drive the R/W head from one track to another track rapidly with bounded control effort, while the main task of track-following is to maintain the R/W head at the center of the target track as closely as possible so that the data can be read from or write to the disk.

Applications of reset control to track-seeking control will be addressed in the subsequent chapters. In the following, we focus on the application of the DF to the track-following control of HDD servo systems. In track-following control, one of the most challenging problems is the mid-frequency disturbance attenuation. By mid-frequency, we mean the frequency range near the gain-crossover frequency of the open loop. As explained in Chapter 1, due to Bode's gain-phase constraint, direct compensation over the mid-frequency by a linear compensator would lead to decrease in phase margin or even instability.

Figure 2.9 Structure of a typical HDD

The mathematical model used for control design is usually a frequency domain model described by the transfer function from the input to the VCM (V) to the position of the R/W head (μm). The transfer function can be obtained by the frequency-response identification method [3]. The transfer function model of the plant considered here is given by

$$Y(s) = P(s)U(s) \tag{2.47}$$

where $Y(s)$ and $U(s)$ are, respectively, the Laplace transforms of the input voltage $u(t)$ to the VCM and the position of the R/W head $y(t)$ with

$$P(s) = \frac{2.09 \times 10^7}{s^2 + 150.8s + 6.317 \times 10^4} \cdot \frac{1.238 \times 10^9}{s^2 + 7037s + 1.238 \times 10^9}.$$

The Bode plot of this transfer function is given in Figure 2.10. The first part of the model characterizes the low-frequency dynamics while the second part is related to the high-frequency dynamics which come from the flexibility of the structure or nonlinearities, etc. A preliminary PI controller is first designed as

$$C(s) = \frac{12.2014(s + 2651)(s + 0.6283)}{s(s + 3.977 \times 10^4)} \tag{2.48}$$

which leads to a gain margin of 4.07 dB at frequency 3.2×10^4 rad/s and a phase margin of 58.2° at frequency 7.03×10^3 rad/s. See Figure 2.10 for the Bode plots of the open-loop transfer function $P(s)C(s)$.

In the following, we consider using an RNBC to provide extra compensation near the gain-crossover frequency. The main idea is depicted in Figure 2.11 where an

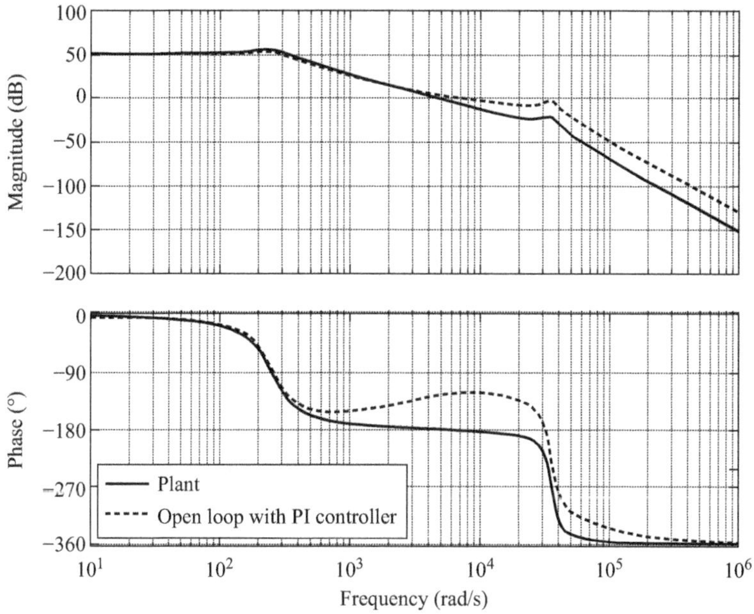

Figure 2.10 The Bode plots of the plant and the open loop with the PI controller

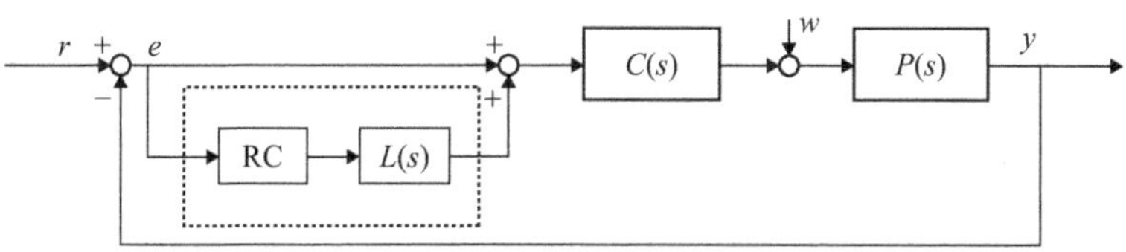

Figure 2.11 Mid-frequency disturbance rejection

RNBC is inserted in parallel. The linear NBC $L(s)$ is used to increase the open-loop gain at mid-frequency range while the reset phase lead controller (RC) is adopted to compensate for the phase lag caused by the NBC. The parallel interconnection was first used in Reference 4 for the traditional reset control design ($R = 0$). The benefit of this interconnection is that the reset action will not decrease the low-frequency open-loop gain. We focus on how the reset matrix R affects the performance of the overall system. The parameters of the NBC and the reset phase lead controller are chosen as

$$\xi_1 = 1, \quad \xi_2 = 1 \times 10^{-4}, \quad f = 800\,\text{Hz},$$

$$f_r = 0.01\,\text{Hz}, \quad k_p = 0.617, \quad k_i = 0.001k_p.$$

Further, let reset matrix R be a free parameter which can be adjusted in simulation.

The ratio ξ_1/ξ_2 determines the phase lag caused by the NBC. When this ratio is small, the phase lag caused is not significant in which case the traditional reset

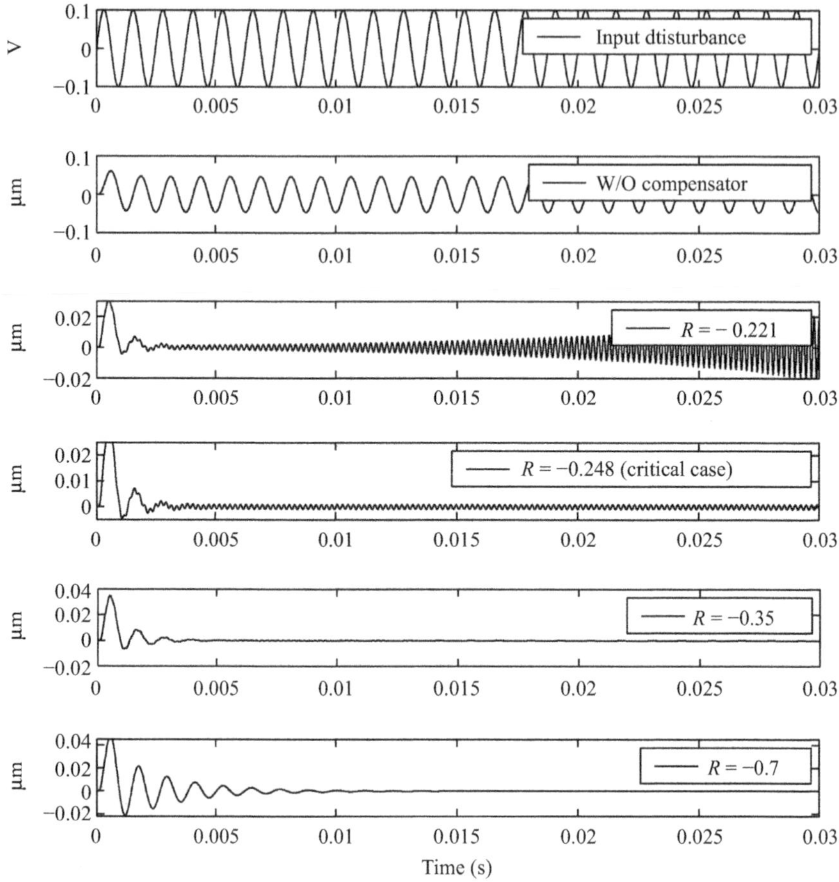

Figure 2.12 Responses to sinusoid disturbance at 800 Hz

scheme (i.e., $R = 0$) is sufficient to compensate for the phase lag. Here we use a large ratio ξ_1/ξ_2 to provide a very large open-loop gain near the gain-crossover frequency. In this case, a zero reset matrix is not sufficient to maintain the stability of the overall system. See the simulation results in the next subsection.

2.3.3 Simulation results

The sampling period of simulation is set to be 2.5×10^{-5} s and the saturation level of the control input is set to 3 V, i.e., $|u| \leq 3$ V:

1. Under the parameters chosen above, the critical value of R for the system to be stable or unstable is about $R = -0.248$. In other words, the closed-loop system is stable only when $R \in [-1, -0.248]$. From this we can see that the traditional reset control (i.e., $R = 0$) is not able to maintain the stability of the overall system.
2. Figure 2.12 gives the output responses for sinusoid disturbance at 800 Hz with deferent values of $R \in [-1, -0.248]$. When $R \in (-1, -0.248)$, the effect of

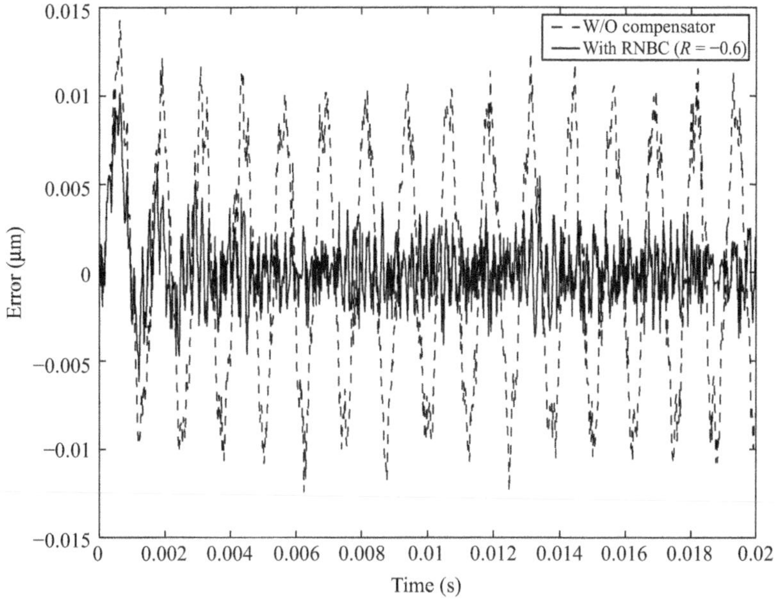

Figure 2.13 Responses to sinusoid disturbance at 800 Hz (with input disturbance, output disturbance and white noises inserted)

the disturbance is almost removed. We also find that when R decreases and approaches -1, the convergence rate of the output would slow down. Actually, as commented in Remark 2.3, this convergence rate is related to $|\lambda(Re^{\frac{\pi}{\omega}A})|$. For fixed ω, the closer the parameter R to -1, the slower the convergence rate. However, we need R to be close to -1 to compensate for the phase lag and maintain the stability. Thus a trade-off has to be made between phase margin (stability) and convergence rate. In our example, a value of R which is close to -0.35 provides nice performance.

In Figure 2.13, input disturbance, output disturbance, and measurement noise are added to simulate the real environment. They are generated by passing normally distributed random signals through disturbance models. In addition, the system suffers from an input disturbance at 800 Hz with magnitude of 0.2 V. We found that when the disturbance and noise are inserted, we need smaller R to achieve better performance. In Figure 2.13, R is set to -0.6. From this figure, we can see that in this case the proposed RNBC is still effective and the disturbance is significantly attenuated.

3. Table 2.1 gives the disturbance attenuation for disturbance frequencies close to the central frequency 800 Hz. Both the cases with and without RNBC are given for comparison. In this simulation, R is set to -0.3 and the amplitude of input disturbance is set to 1 V.

We point out that the reduction rates in Table 2.1 are not sensitive to the amplitude of the input signal, though the reset system is nonlinear. This phenomenon can be

Table 2.1 The responses to disturbance near central frequency $f = 800$ Hz. (f_d: disturbance frequency, A_1: error amplitude without RNBC, and A_2: error amplitude with RNBC)

f_d (Hz)	A_1 (μm)	A_2 (μm)	Reduction rate $\frac{A_1 - A_2}{A_1}$%
500	0.674	0.372	44.80%
600	0.5885	0.2725	53.70%
700	0.52	0.1575	69.71%
800	0.462	0.0014	99.70%
900	0.416	0.192	53.85%
1000	0.377	0.307	18.57%
1100	0.344	0.326	5.23%
1200	0.316	0.311	1.58%

interpreted by the fact that the DF of a reset system is independent of the amplitude of the input signal. See Theorem 2.1.

The above simulations show that the proposed compensator can provide large open-loop gain around the gain-crossover frequency without significant degradation of stability. They also show that an unstable base linear system can be stabilized by applying reset. When $R = 1$ (i.e., without reset), the closed loop is unstable due to phase margin loss. When $R \in [-1, -0.248)$, the simulation results show that the system is asymptotically stable. This observation has also been made in Reference 2.

Notes

This chapter is mainly based on References 5 and 6. Early literature about RCSs include References 7, 8, and 9 where the DFs of the CI and the FORE were introduced. DF analysis of general reset systems with zero reset matrices can be found in References 4 and 1. The parallel way of the insertion of the RNBC depicted in Figure 2.11 and the HDD model used in this chapter come from Reference 4 which focuses on reset systems with zero reset matrices. Example 2.2 was used in Reference 2 to explain the benefit of reset control in the time domain. Recent developments of DF-based control design can be found in References 10 and 11 where the concept of reset band is introduced. For comprehensive introductions of the modeling and control of HDD servo systems, we refer the reader to References 12 and 13.

References

[1] Y. Li, G. Guo, and Y. Wang. Reset control for midfrequency narrowband disturbance rejection with an application in hard disk drives. *IEEE Transactions on Control Systems Technology*, 19(6):1339–1348, 2011.

[2] O. Beker, C.V. Hollot, Y. Chait, and H. Han. Fundamental properties of reset control systems. *Automatica*, 40(6):905–915, 2004.

[3] P. Eykhoff. *System identification parameter and state estimation*. John Wiley, New York, NY, 1974.

[4] Y. Li, G. Guo, and Y. Wang. Nonlinear mid-frequency disturbance compensation in hard disk drives. In *Proceedings of the 16th International Federation Control Conference*, Elsevier Science, Prague, pp. 3–8, 2005.

[5] Y. Guo, Y. Wang, and L. Xie. Frequency-domain properties of reset systems with application in hard-disk-drive systems. *IEEE Transactions on Control Systems Technology*, 17(6):1446–1453, 2009.

[6] Y. Guo, Y. Wang, and L. Xie. Mid-frequency disturbance rejection of HDD systems. In *Chinese Control Conference.* July 26–31, 2007, pp. 56–60. IEEE, Hunan, China, 2007.

[7] J.C. Clegg. A nonlinear integrator for servomechanisms. *Transactions of the American Institute of Electrical Engineers*, 77(Part II):41–42, 1958.

[8] K.R. Krishnan and I.M. Horowitz. Synthesis of a nonlinear feedback system with significant plant-ignorance for prescribed system tolerances. *International Journal of Control*, 19(4):689–706, 1974.

[9] I. Horowitz and P. Rosenbaum. Nonlinear design for cost of feedback reduction in systems with large parameter uncertainty. *International Journal of Control*, 21(6):977–1001, 1975.

[10] A. Baños, S. Dormido, and A. Barreiro. Limit cycles analysis of reset control systems with reset band. *Nonlinear Analysis: Hybrid Systems*, 5(2):163–173, 2011.

[11] A. Barreiro, A. Baños, S. Dormido, and J.A. González-Prieto. Reset control systems with reset band: well-posedness, limit cycles and stability analysis. *Systems & Control Letters*, 63:1–11, 2014.

[12] B.M. Chen, T.H. Lee, K. Peng, and V. Venkataramanan. Hard disk drive servo systems. In *Advances in Industrial Control*, Springer-Verlag London Ltd., 2006.

[13] A. Al Mamun, G.X. Guo, and C. Bi. Hard disk drive: mechatronics and control. *In Automation and Control Engineering*, vol. 23. CRC, Taylor & Francis Group, 2006.

Chapter 3
Stability of reset control systems

Stability is a basic requirement for *reset control systems* (RCSs). However, for RCSs with zero-crossing type of reset, because of the dependence of the reset time instants on the system state and input, the stability analysis of RCSs is challenging. This chapter presents some major developments concerning the stability of RCSs with zero-crossing type of reset in recent years. These results include quadratic stability of RCSs with and without time-delays, reset time-dependent stability criteria, and finite $\mathcal{L}_2$-gain stability. These results were originally developed for RCSs with zero reset matrices, i.e., part of the state is reset to zero whenever the tracking error crosses zero, but most of them are slightly modified to adapt to arbitrarily reset matrices case in this chapter.

3.1 Preliminaries

3.1.1 Annihilator of matrices

One of the tools of this chapter and the following Chapter 4 is the notion of full-rank right annihilator of a matrix defined in the following.

Definition 3.1. *Let C be any constant $m \times n$ $(m \leq n)$ matrix with* Rank $C = r_c$. *A full-rank right annihilator of C is an $n \times (n-r)$ matrix, denoted by $C^{\perp}$, satisfying:*

1. $CC^{\perp} = 0$;
2. *$C^{\perp}$ is of full column rank, i.e.,* Rank $(C^{\perp}) = n - r$.

A full-rank right annihilator of C is actually a matrix whose columns constitute a set of bases for ker (C), the *kernel space* of C, i.e.,

$$\ker(C) = \mathrm{Im}(C^{\perp}). \tag{3.1}$$

Note that the full-rank right annihilator of a given matrix is not unique. In general, we can solve the linear equations $Cx = 0$ for a set of $n - r$ linear independent solutions in order to get $C^{\perp}$. For some simple matrices, it can be directly constructed. For instance, for the n-dimensional row vector

$$C = [1\ 0\ \cdots\ 0],$$

we can construct a full-rank right annihilator of C as

$$C^{\perp} = \begin{bmatrix} 0_{1\times(n-r)} \\ I_{n-r} \end{bmatrix}.$$

The following lemma is obviously true.

Lemma 3.1. *Suppose that C is an $m \times n$ matrix and $C^{\perp}$ is one of the full-rank right annihilators of C. Then the following claims hold:*

1. $x \in \ker(C)$ *if and only if there is an* $\eta \in \mathbb{R}^{n-r}$ *such that*

   $$x = C^{\perp}\eta;$$

2. *For any $n \times q$ matrix P,*

   $$CP = 0$$

   if and only if there is an $(n-r) \times q$ matrix Γ such that

   $$P = C^{\perp}\Gamma.$$

3.1.2 Passive systems

For any function $f : \mathbb{R}_{\geq 0} \to \mathbb{R}$ and any $T \geq 0$, define the *truncated function*

$$f_T(t) = \begin{cases} f(t), & t \leq T \\ 0, & t > 0. \end{cases}$$

Denote by $\mathcal{L}_2$ the set of measurable functions $f : \mathbb{R}_{\geq 0} \to \mathbf{R}$ such that

$$\int_0^{\infty} |f(s)|ds < \infty.$$

Define $\mathcal{L}_{2,e}$ as

$$\mathcal{L}_{2,e} := \{f \mid f_T \in \mathcal{L}_2, \ \forall T \geq 0\}.$$

For any $f = (f_1, f_2, \cdots, f_n)^T$ with $f_i \in \mathcal{L}_2$, the $\mathcal{L}_2$-*norm* of f is defined as

$$\|f\| := \left[\int_0^{\infty} f^T(s)f(s)ds\right]^{1/2}.$$

In this section, we briefly review some basic concepts and results on passivity of the input-output system defined as an operator $H : \mathcal{L}_{2,e} \to \mathcal{L}_{2,e}$, i.e.,

$$\Sigma_{io}: \quad y = H(u), \quad u \in \mathcal{L}_{2,e}. \tag{3.2}$$

When dissipativeness is considered, we focus on the kind of systems $H : \mathcal{L}_{2,e} \to \mathcal{L}_{2,e}$ which has a state-space representation

$$\Sigma_{ss}: \quad \begin{cases} \dot{x} = f(x, u) \\ y = g(x, u) \end{cases} \tag{3.3}$$

where $x \in \mathbb{R}^n$ is the state variable, $u, y \in \mathcal{L}_{2,e}$ are the input and output, respectively.

Definition 3.2. *[1, 2] A system $y = H(u)$ is said to be finite $\mathcal{L}_2$-gain stable if there is a constant $\gamma > 0$ such that*

$$\|y_T\| \leq \gamma \|u_T\|, \quad \forall T \geq 0. \tag{3.4}$$

Definition 3.3. *[1, 2] A system $y = H(u)$ is called passive if there is a constant β such that*

$$\int_0^T y^T(s)u(s)ds \geq \beta, \quad \forall t \geq 0, \forall u \in \mathcal{L}_{2,e}. \tag{3.5}$$

If, in addition, there are constants $\delta \geq 0$, $\varepsilon \geq 0$ such that

$$\int_0^T y^T(s)u(s)ds \geq \beta + \delta \|u_T\|^2 + \varepsilon \|y_T\|^2, \quad \forall T \geq 0, \forall u \in \mathcal{L}_{2,e}, \tag{3.6}$$

then the system is, respectively, said to be:

1. *input strictly passive if $\delta > 0$;*
2. *output strictly passive if $\varepsilon > 0$;*
3. *very strictly passive if $\delta > 0$, $\varepsilon > 0$.*

The system is said to be pseudo very strictly passive if there are constants β, δ, $\varepsilon \in \mathbb{R}$ such that (3.6) holds (δ and ε are not required to be nonnegative).

Theorem 3.1. *[2] System $y = H(u)$ is passive if there is a continuous function $V(t) \geq 0$ such that*

$$V(T) - V(0) \leq \int_0^T y^T(s)u(s)ds, \quad \forall T \geq 0, \forall u \in \mathcal{L}_{2,e}, \forall V(0). \tag{3.7}$$

In addition, assume that there are constants $\delta \geq 0$, $\varepsilon \geq 0$ such that

$$V(T) - V(0) \leq \int_0^T y^T(s)u(s)ds - \delta \|u_T\|^2 - \varepsilon \|y_T\|^2,$$
$$\forall T \geq 0, \forall u \in \mathcal{L}_{2,e}, \forall V(0). \tag{3.8}$$

Then the system is input strictly passive if there is a $\delta > 0$, output strictly passive if there is a $\varepsilon > 0$, and very strictly passive if there is a $\delta > 0$ and $\varepsilon > 0$ such that the inequality (3.8) holds.

Definition 3.4. *(**Dissipativeness** [2]) System (3.3) is said to be dissipative with respect to the supply rate $w(u, y)$ if there exists a so-called storage function $V(x) \geq 0$ such that*

$$V(x(T)) \leq V(x(0)) + \int_0^T w(u(s), y(s))ds, \quad \forall T \geq 0, \forall u \in \mathcal{L}_{2,e}, \forall x(0) \tag{3.9}$$

Remark 3.1. *System (3.3) is: (1) passive; (2) input strictly passive; (3) output strictly passive; and (4) very strictly passive, respectively, if it is dissipative with respect to the supply rate:*

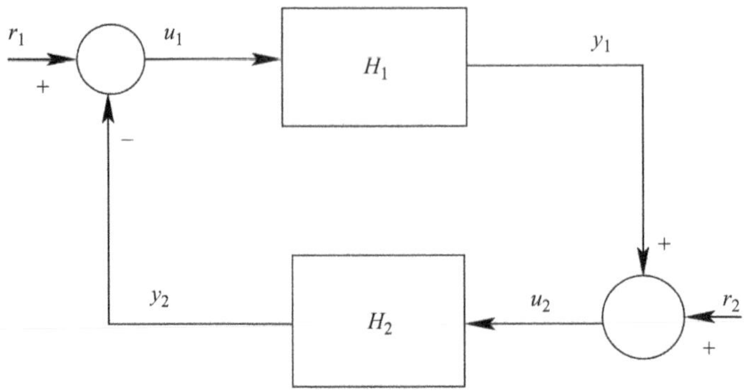

Figure 3.1 Negative feedback interconnection of H_1 and H_2

1. $w_p(u,y) = u^T y$;
2. $w_i(u,y) = u^T y - \delta u^T u,\ \delta > 0$;
3. $w_o(u,y) = u^T y - \varepsilon y^T y,\ \varepsilon > 0$; *and*
4. $w_v(u,y) = u^T y - \delta u^T u - \varepsilon y^T y,\ \delta > 0,\ \varepsilon > 0,$

respectively. See References 1 and 2.

In the following, we consider the closed-loop system consisting of two systems

$$y_1 = H_1(u_1), \quad y_2 = H_2(u_2)$$

interconnected via negative feedback

$$u_1 = r_1 - y_2, \quad u_2 = r_2 + y_1.$$

The input and output of this closed-loop system are, respectively, $r = (r_1,\ r_2)^T$ and $y = (y_1,\ y_2)^T$. We refer to this feedback system as $\Sigma_{fb}(H_1,\ H_2)$ which is depicted in Figure 3.1.

Theorem 3.2. *(**Passivity Theorem** [3, 2]) The feedback system $\Sigma_{fb}(H_1,\ H_2)$ is finite $\mathcal{L}_2$-gain stable if the following two conditions hold:*

1. *both H_1 and H_2 are pseudo very strictly passive with*

$$\int_0^T y_1^T(s)u_1(s)ds \geq \beta_1 + \delta_1\|u_{1T}\|^2 + \varepsilon_1\|y_{1T}\|^2, \quad \forall T \geq 0, \forall u \in \mathcal{L}_{2,e},$$

$$\int_0^T y_2^T(s)u_2(s)ds \geq \beta_2 + \delta_2\|u_{2T}\|^2 + \varepsilon_2\|y_{2T}\|^2, \quad \forall T \geq 0, \forall u_2 \in \mathcal{L}_{2,e};$$

2. *there hold that*

$$\delta_1 + \varepsilon_2 > 0, \quad \delta_2 + \varepsilon_1 > 0.$$

Corollary 3.1. *[2] The feedback system $\Sigma_{fb}(H_1,\ H_2)$ is finite $\mathcal{L}_2$-gain stable if one of the following conditions holds:*

1. *both H_1 and H_2 are input strictly passive ($\varepsilon_1, \varepsilon_2 > 0, \delta_1 = \delta_2 = 0$);*
2. *both H_1 and H_2 are output strictly passive ($\delta_1, \delta_2 > 0, \varepsilon_1 = \varepsilon_2 = 0$);*
3. *H_1 is very strictly passive and H_2 is passive ($\varepsilon_1, \delta_1 > 0, \delta_2 = \varepsilon_2 = 0$);*
4. *H_1 is passive and H_2 is very strictly passive ($\varepsilon_2, \delta_2 > 0, \delta_1 = \varepsilon_1 = 0$).*

In the following, we introduce an important class of passive systems, namely *linear time-invariant* (LTI) systems whose transfer functions are *positive real.*

Definition 3.5. *[2] A $p \times p$ proper rational transfer function matrix $G(s)$ is called positive real if*

- *poles of all elements of $G(s)$ are with nonpositive real parts;*
- *for all real ω such that $j\omega$ is not a pole of any element of $G(s)$, the matrix*

  $$G(j\omega) + G^T(j\omega)$$

  is positive semi-definite;
- *any pure imaginary pole $j\omega$ of any element of $G(s)$ is a simple pole and the residue matrix $\lim_{s\to j\omega}(s - j\omega)G(s)$ is positive semi-definite Hermitian.*

The transfer function $G(s)$ is called strictly positive real if $G(s - \varepsilon)$ is positive real for some $\varepsilon > 0$.

Remark 3.2. *For a scalar transfer function, the second condition of the above definition degenerates to*

$$\mathrm{Re}\,[G(j\omega)] \geq 0, \quad \forall \omega \in \mathbb{R},$$

i.e., the Nyquist plot of $G(s)$ lies in the closed right-half complex plane.

The following theorems give some relationships between passivity and positive realness.

Theorem 3.3. *[2] A LTI system is passive if and only if its transfer function $h(s)$ is positive real.*

Theorem 3.4. *[2] Suppose a LTI system (A, B, C, D) is strictly positive real and there are matrices $P = P^T > 0$, L, W such that*

$$\begin{bmatrix} -PA - A^TP + L^TL & L^TW \\ W^TL & D + D^T \end{bmatrix} > 0. \tag{3.10}$$

Then the system is very strictly passive.

Theorem 3.5. *[2] Given a LTI system with a rational transfer function matrix $H(s)$. Assume that all the poles of $H(s)$ are with negative real parts. Then the following hold:*

1. *The system is passive if and only if*

   $$\lambda[H(j\omega) + H^*(j\omega)] \geq 0, \quad \omega \in [-\infty, +\infty];$$

2. *The system is input strictly passive if and only if there is a constant* $\delta > 0$ *such that*

$$\lambda[H(j\omega) + H^*(j\omega)] \geq \delta > 0, \quad \omega \in [-\infty, +\infty].$$

In addition, if the system is single-input single-output with transfer function h(s), then it is output strictly passive if and only if there is a constant $\varepsilon > 0$ *such that*

$$\mathrm{Re}\,[h(j\omega)] \geq \varepsilon |h(j\omega)|^2.$$

Lemma 3.2. *(***Kalman-Yakubovich-Popov Lemma** *[2]) Let* $G(s) = C(sI - A)^{-1}B + D$ *be a* $p \times p$ *transfer function matrix where (A, B) is controllable and (A, C) is observable. Then the following statements hold:*

1. *G(s) is positive real if and only if there exist matrices* $P = P^T > 0$*, L, and W such that*

$$\begin{aligned} PA + A^T P &= -L^T L \\ PB &= C^T - L^T W \\ W^T W &= D + D^T. \end{aligned} \tag{3.11}$$

2. *G(s) is strictly positive real if and only if there exist matrices* $P = P^T > 0$*, L, and W and a constant* $\varepsilon > 0$ *such that*

$$\begin{aligned} PA + A^T P &= -L^T L - \varepsilon P \\ PB &= C^T - L^T W \\ W^T W &= D + D^T. \end{aligned} \tag{3.12}$$

Lemma 3.3. *[2] Given* $A \in \mathbb{R}^{n\times n}$, $B \in \mathbb{R}^{n\times m}$, $M = M^T \in \mathbb{R}^{(n+m)\times(n+m)}$*. Suppose that (A, B) is controllable and*

$$\det(j\omega I - A) \neq 0, \quad \forall \omega \in \mathbb{R}.$$

Then the following two statements are equivalent:

1. *For all* $\omega \in \mathbb{R} \cup \{\infty\}$*, there holds*

$$\begin{bmatrix} (j\omega I - A)^{-1}B \\ I \end{bmatrix}^* M \begin{bmatrix} (j\omega I - A)^{-1}B \\ I \end{bmatrix} \leq 0.$$

2. *There exists a matrix* $P = P^T \in \mathbb{R}^{n\times n}$ *such that*

$$M + \begin{bmatrix} A^T P + PA & PB \\ B^T P & 0 \end{bmatrix} \leq 0.$$

The corresponding equivalence for strict inequalities holds even if (A, B) is not controllable.

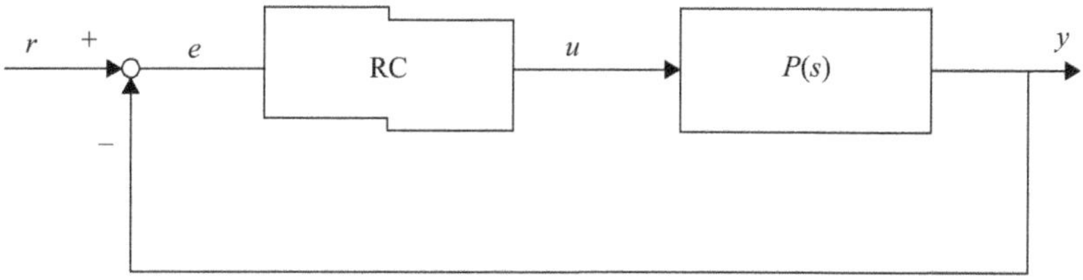

Figure 3.2 Reset control system

3.2 Quadratic stability

Consider an RCS depicted in Figure 3.2 where the plant $P(s)$ is given by

$$\Sigma_p : \quad \begin{cases} \dot{x}_p = A_p x_p + B_p u \\ y = C_p x_p \end{cases} \tag{3.13}$$

with $x \in \mathbb{R}^{n_p}$, $u \in \mathbb{R}^m$, $y \in \mathbb{R}^1$, and RC is the reset controller given by

$$\Sigma_R : \quad \begin{cases} \dot{x}_r = A_r x_r + B_r e, & e \neq 0 \\ x_r^+ = R_r x_r, & e = 0 \\ u = C_r x_r \end{cases} \tag{3.14}$$

with $x_r \in \mathbb{R}^{n_r}$. Combining (3.13) with (3.14) gives

$$\begin{cases} \dot{x} = Ax + Br, & x \notin \mathcal{J} \\ x^+ = Rx, & x \in \mathcal{J}, \end{cases} \tag{3.15}$$

where $x = \begin{bmatrix} x_p \\ x_r \end{bmatrix} \in \mathbb{R}^n$ with $n = n_p + n_r$,

$$A = \begin{bmatrix} A_p & B_p C_r \\ -B_r C_p & A_r \end{bmatrix}, \quad R = \begin{bmatrix} I & \\ & R_r \end{bmatrix}, \quad B = \begin{bmatrix} 0 \\ B_r \end{bmatrix}, \quad C = [C_p \ \ 0]$$

and

$$\mathcal{J} := \{x \in \mathbb{R}^n \mid r - Cx = 0\}. \tag{3.16}$$

The reset matrix R_r is assumed to be of the block-diagonal form

$$R_r = \begin{bmatrix} I_{n_r - \rho} & \\ & R_\rho \end{bmatrix}$$

where R_ρ is a $\rho \times \rho$ matrix with $\rho \leq n_r$ and $I_{n_r-\rho}$ is the identity matrix with dimension $n_r - \rho$. This structure indicates that part of the controller states are reset.

Note that the jump set $\mathcal{J}$ defined in (3.16) may lead to Zeno behaviors. In order to avoid them and ensure well-posedness, we can redefine the jump set by removing the invariant subset of the reset matrix R from $\mathcal{J}$. For instance, in the case with $R_\rho = 0_{\rho\times\rho}$, the jump set can be defined as

$$\mathcal{J} := \{x \in \mathbb{R}^n \mid r - Cx = 0,\ (I - R)x \neq 0\}. \tag{3.17}$$

Another way to avoid Zeno behaviors is to impose time-regularization [4, 5] on the reset rate, i.e., the reset action is triggered only when the time spent since the latest reset action is greater than a pre-specified positive number $\tau_{\min}$. Thus, there are only a finite number of reset actions during any compact time interval.

For stability analysis, we set $r = 0$ in the following.

Theorem 3.6. *[6] If there exists a continuously differentiable, positive definite, and radially unbounded function $V(x) : \mathbb{R}^{n_p+n_c} \to \mathbb{R}$ such that*

$$\dot V(x) < 0, \quad x \neq 0 \tag{3.18}$$

$$\Delta V(x) := V(Rx) - V(x) \le 0, \quad x \in \ker C, \tag{3.19}$$

then:

1. *for any initial condition, system (3.15) has a solution which is well-defined for all $t \ge 0$;*
2. *system (3.15) with $r = 0$ is globally asymptotically stable.*

Remark 3.3. *The conditions in Theorem 3.6 require that the baseline system*

$$\dot x = Ax \tag{3.20}$$

be asymptotically stable. However, it is worth pointing out that a stable reset system might have an unstable baseline system. For instance, consider the planar reset system (3.15) with

$$A = \begin{pmatrix} \alpha & \beta \\ -\beta & \alpha \end{pmatrix}, \quad C = \begin{pmatrix} 1 & -1/k \end{pmatrix}, \quad R = \begin{pmatrix} 1 & 0 \\ 0 & 0 \end{pmatrix}$$

where α, β, and k are positive numbers. The baseline system $\dot x = Ax$ is unstable since $\alpha > 0$. It is easy to check that the reset interval is constant

$$\Delta = \frac{\pi - \arctan k}{\beta}.$$

Thus this reset system is globally asymptotically stable if and only if $A\exp(A\Delta)$ is Schur stable, or equivalently,

$$\frac{1}{1+k^2}\exp\left[\frac{[2\alpha(\pi - \arctan k)}{\beta}\right] < 1. \tag{3.21}$$

The phase trajectory of the planar reset system with $\alpha = 4$, $\beta = 10$, $k = 3$ and, initial state $(10, 5)$ is given in Figure 3.3.

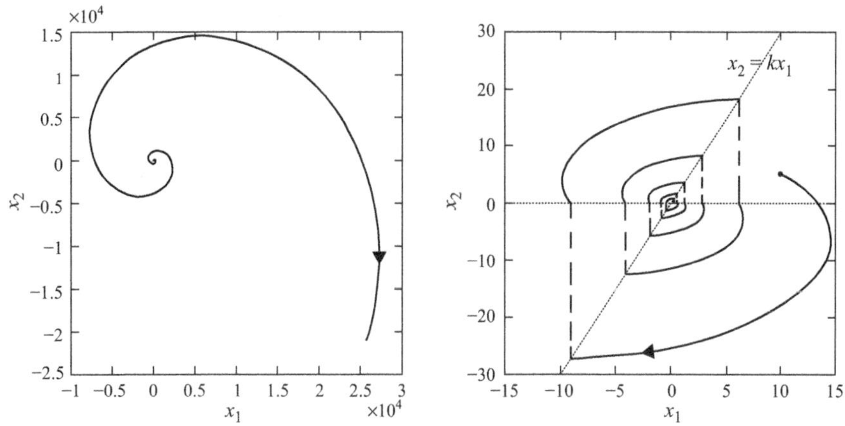

Figure 3.3 Phase trajectory of an RCS with an unstable baseline system. (Left: baseline system; right: reset system)

Definition 3.6. *RCS (3.15) is said to be quadratically stable if the conditions of Theorem 3.6 hold with*

$$V(x) = x^T Px \tag{3.22}$$

where P is a positive-definite symmetric matrix. That is, there is a $P > 0$ such that

$$A^T P + PA < 0, \tag{3.23}$$

$$x^T (R^T PR - P)x \leq 0, \quad \forall x \in \ker C. \tag{3.24}$$

Theorem 3.7. *Suppose that the reset matrix R_ρ is Schur stable, i.e., $|\lambda(R_\rho)| < 1$. Then RCS (3.15) is quadratically stable if and only if there are matrices $P_{11} = P_{11}^T \in \mathbb{R}^{n_p \times n_p}$, $P_{12} \in \mathbb{R}^{n_p \times (n_r - \rho)}$, $P_{22} = P_{22}^T \in \mathbb{R}^{(n_r-\rho)\times(n_r-\rho)}$, $P_{33} = P_{33}^T \in \mathbb{R}^{\rho \times \rho}$ and a vector $\beta \in \mathbb{R}^\rho$ such that*

$$P = \begin{bmatrix} P_{11} & P_{12} & C_p^T \beta^T \\ \star & P_{22} & 0_{(n_r-\rho)\times\rho} \\ \star & \star & P_{33} \end{bmatrix} > 0 \tag{3.25}$$

$$A^T P + PA < 0 \tag{3.26}$$

$$R_\rho^T P_{33} R_\rho - P_{33} \leq 0. \tag{3.27}$$

Proof: According to the definition, the quadratic stability of the system is equivalent to the existence of a positive-definite matrix P such that (3.23) and (3.24) hold. We only need to prove that (3.24) is equivalent to (3.25) and (3.27). Note that by Lemma 3.1, (3.24) holds if and only if

$$(C^\perp)^T (R^T PR - P)C^\perp \leq 0. \tag{3.28}$$

Suppose that $C_p^\perp$ is a full-rank right annihilator of C_p. By the structure of C, $C^\perp$ can be chosen as

$$C^\perp = \begin{bmatrix} C_p^\perp & & \\ & I_{n_r-\rho} & \\ & & I_\rho \end{bmatrix}. \tag{3.29}$$

Partition P and R as

$$P = \begin{bmatrix} P_{11} & P_{12} & P_{13} \\ \star & P_{22} & P_{23} \\ \star & \star & P_{33} \end{bmatrix}, \quad R = \begin{bmatrix} I_{n_p} & & \\ & I_{n_r-\rho} & \\ & & R_\rho \end{bmatrix}. \tag{3.30}$$

By substituting (3.29) and (3.30) into (3.28), we get

$$\begin{bmatrix} 0 & 0 & (C_p^\perp)^T P_{13}(R_\rho - I) \\ \star & 0 & P_{23}(R_\rho - I) \\ \star & \star & R_\rho^T P_{33} R_\rho - P_{33} \end{bmatrix} \le 0, \tag{3.31}$$

which is equivalent to

$$\begin{cases} (C_p^\perp)^T P_{13}(R_\rho - I) = 0 \\ P_{23}(R_\rho - I) = 0 \\ R_\rho^T P_{33} R_\rho - P_{33} \le 0. \end{cases} \tag{3.32}$$

By the assumption that R_ρ is Schur stable, the first two equations in (3.33) are equivalent to

$$(C_p^\perp)^T P_{13} = 0, \quad P_{23} = 0.$$

$(C_p^\perp)^T P_{13} = 0$ is equivalent to $P_{13}^T C_p^\perp = 0$, which is in turn equivalent to the existence of a vector $\beta \in \mathbb{R}^\rho$ such that $P_{13}^T = \beta C_p$. Thus (3.28) holds if and only if

$$\begin{cases} P_{13} = C_p^T \beta^T \\ P_{23} = 0 \\ R_\rho^T P_{33} R_\rho - P_{33} \le 0 \end{cases} \tag{3.33}$$

hold for some vector β.

Corollary 3.2. *Planar RCS (3.15) with $n_p = n_r = n_\rho = 1$ and $C_p \neq 0$ is quadratically stable if and only if*

$$\operatorname{Re} \lambda(A) < 0, \quad |R_\rho| \le 1.$$

Proof: By using Theorem 3.7, the RCS is quadratically stable if and only if there are constants $P_{11} \in \mathbb{R}^1$, $P_{33} = P_{33}^T \in \mathbb{R}^1$, and $\beta \in \mathbb{R}^1$ such that

$$P = \begin{bmatrix} P_{11} & C_p^T \beta^T \\ \star & P_{33} \end{bmatrix} > 0$$

and (3.26) and (3.27) hold. Since C_p is nonzero, the reset action does not impose any structural constraint on the matrix P. In addition, (3.27) holds if and only if $|R_\rho| \leq 1$ so long as $P_{33} > 0$, and there is a positive matrix P such that (3.26) holds if and only if A is stable, i.e., $\mathrm{Re}\,\lambda(A) < 0$.

Theorem 3.8. *(H_β-**condition** [6]) Suppose that the reset matrix $R_\rho = 0_{\rho\times\rho}$. Then RCS (3.15) is quadratically stable if and only if the H_β-condition holds, i.e., there exist a vector $\beta \in \mathbb{R}^\rho$ and a positive definition matrix $P_\rho \in \mathbb{R}^{\rho\times\rho}$ such that the transfer function*

$$H_\beta(s) := [\beta C_p \quad 0_{\rho\times(n_r-\rho)} \quad P_\rho](sI - A)^{-1}\begin{bmatrix} 0_{(n-\rho)\times\rho} \\ I_\rho \end{bmatrix} \tag{3.34}$$

is strictly positive real.

Proof: According to the proof of Theorem 3.7, if $R_\rho = 0_{\rho\times\rho}$, then the last inequality of (3.33) naturally holds since P is assumed to be positive definite, and (3.28) holds if and only if there is a vector $\beta \in \mathbb{R}^\rho$ such that

$$\begin{cases} P_{13} = C_p^T\beta^T \\ P_{23} = 0. \end{cases} \tag{3.35}$$

Thus (3.28) holds for some positive matrix $P \in \mathbb{R}^n$ if and only if there exist a positive matrix $P_\rho \in \mathbb{R}^\rho$ and a vector $\beta \in \mathbb{R}^\rho$ such that

$$[0_{\rho\times(n-\rho)} \quad I_\rho]P = [\beta C_p \quad 0_{\rho\times(n_r-\rho)} \quad P_\rho]. \tag{3.36}$$

By using Lemma 3.2, the RCS (3.15) is quadratically stable if and only if $H_\beta(s)$ is strictly positive real.

Remark 3.4. *From the proof of Theorems 3.7 and 3.8, the H_β-condition is equivalent to the solvability of the linear matrix inequality (LMIs) (3.25) and (3.26). Thus if the H_β-condition holds for a positive matrix P_ρ, then the RCS is quadratically stable for any reset matrix R_ρ satisfying*

$$R_\rho^T P_\rho R_\rho - P_\rho \leq 0.$$

The following corollary shows that when the last component of the system state is the only state to be reset, then the quadratic stability of the system can be checked by a single-variable search.

Corollary 3.3. *[6] RCS (3.15) with $n_\rho = 1$ is quadratically stable if and only if $|R_\rho| \leq 1$ and there exists a real number $\beta \in \mathbb{R}^1$ such that the transfer function*

$$H_\beta(s) := \begin{bmatrix} \beta C_1 & 1 \end{bmatrix}(sI - A)^{-1}\begin{bmatrix} 0_{(n-1)\times 1} \\ 1 \end{bmatrix} \tag{3.37}$$

is strictly positive real, where $C_1 = [C_p \quad 0_{1\times(n_r-1)}]$.

Example 3.1. *Consider the integrator*

$$\dot{x}_p = u_p$$

controlled by a first-order reset element (FORE)

$$\begin{cases} \dot{x}_r = -x_r + e, & e \neq 0 \\ x_r^+ = R_\rho x_r, & e = 0 \end{cases} \tag{3.38}$$

with $u = x_r$, $e = r - x_p$. *The closed-loop system with* $r = 0$ *is of the form (3.15) with*

$$A = \begin{bmatrix} 0 & 1 \\ -1 & -1 \end{bmatrix}, \quad C = [1 \;\; 0], \quad R = \begin{bmatrix} 1 & 0 \\ 0 & R_\rho \end{bmatrix}.$$

By (3.37), there holds

$$H_\beta(s) = \frac{s + \beta}{s^2 + s + 1}. \tag{3.39}$$

It is easy to check that $H_\beta(s)$ *is a stable transfer function. In addition, for any* $\varepsilon > 0$,

$$\operatorname{Re} H_\beta(j\omega - \varepsilon) = \frac{(\varepsilon^2 - \varepsilon + 1 - \omega^2)(\beta - \varepsilon) + \omega^2(1 - 2\varepsilon)}{(\varepsilon^2 - \varepsilon + 1 - \omega^2)^2 + \omega^2(1 - 2\varepsilon)^2}.$$

Choose any $\varepsilon \in (0, 0.5)$ *and set* $\beta = \varepsilon$, *then there holds*

$$\operatorname{Re} H_\beta(j\omega - \varepsilon) > 0, \quad \forall \omega > 0.$$

Thus this RCS is quadratically stable for any R_ρ *with* $|R_\rho| \leq 1$, *by using Corollary 3.3. When* $\beta = \varepsilon$, *we have*

$$H_\beta(s - \varepsilon) = \frac{s}{s^2 + (1 - 2\beta)s + (1 + \beta^2)}$$

whose Nyquist plots for different choice of β *are depicted in Figure 3.4. It is worth pointing out that the quadratic stability of this RCS follows directly by using Corollary 3.2.*

In the following, we consider a class of RCSs depicted in Figure 3.2 where

$$P(s) = \frac{\omega_n^2(s + b)}{s(s + 2\zeta\omega_n)}$$

and the reset controller is a FORE with a pole at $s = -b$, i.e.,

$$\begin{cases} \dot{x}_r = -b x_r + e, & e \neq 0 \\ x_r^+ = R_r x_r, & e = 0. \end{cases} \tag{3.40}$$

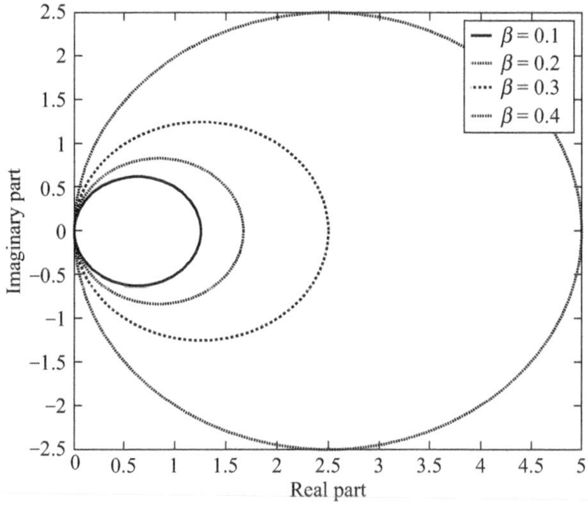

Figure 3.4 The Nyquist plots of $H_\beta(s-\varepsilon)$ for different β

The closed-loop system can then be represented in the form of (3.15) with

$$A = \begin{pmatrix} -2\zeta\omega_n & 1 & 1 \\ 0 & 0 & b \\ -\omega_n^2 & 0 & -b \end{pmatrix}, \quad B = \begin{pmatrix} 0 \\ 0 \\ 1 \end{pmatrix}, \quad R = \begin{pmatrix} 1 & 0 & 0 \\ 0 & 1 & 0 \\ 0 & 0 & R_r \end{pmatrix},$$
$$C = \begin{pmatrix} \omega_n^2 & 0 & 0 \end{pmatrix}. \tag{3.41}$$

Theorem 3.9. *[6] The RCS (3.15)–(3.41) is quadratically stable for all positive b, ω_n, ζ, and all $|R_r| \le 1$.*

3.3 Stability of RCSs with time-delay

Consider an RCS depicted in Figure 3.5 with $r = 0$ and consisting of a plant

$$\Sigma_p: \quad \begin{cases} \dot{x}_p = A_p x_p + B_p u_p \\ y_p = C_p x_p \end{cases} \tag{3.42}$$

with $x \in \mathbb{R}^{n_p}$, $u \in \mathbb{R}^m$, $y \in \mathbb{R}^1$, and a reset controller

$$\Sigma_R: \quad \begin{cases} \dot{x}_r = A_r x_r + B_r u_r, & C_p x_p \ne 0 \\ x_r^+ = R_r x_r, & C_p x_p = 0 \\ y_r = C_r x_r + D_r u_r \end{cases} \tag{3.43}$$

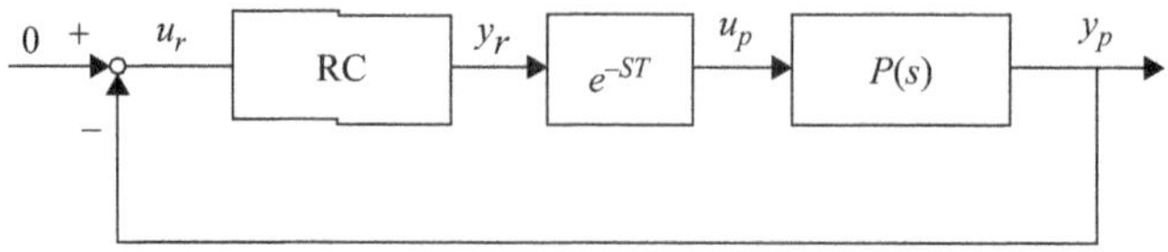

Figure 3.5 RCS with time-delay

with $x_r \in \mathbb{R}^{n_r}$ interconnected via

$$\begin{cases} u_p(t) = y_r(t-\tau) \\ u_r(t) = -y_p(t) \end{cases} \tag{3.44}$$

where $\tau > 0$ is the controller-to-actuator time-delay. The closed-loop system is given by

$$\begin{cases} \dot{x}(t) = Ax(t) + A_d x(t-\tau), & x(t) \notin \ker C \\ x(t^+) = Rx(t), & x(t) \in \ker C \\ x(t) = \phi(t), & t \in [-\tau, 0], \end{cases} \tag{3.45}$$

where $x(t) = \left[x_p^T(t), x_r^T(t)\right]^T \in \mathbb{R}^n$, $n = n_p + n_r$, $C = [C_p \ \ 0]$ and

$$A = \begin{bmatrix} A_p & 0 \\ -B_r C_p & A_r \end{bmatrix}, \quad A_d = \begin{bmatrix} -B_p D_r C_p & B_p C_r \\ 0 & 0 \end{bmatrix}, \quad R = \begin{bmatrix} I_{n_p} & \\ & R_r \end{bmatrix}.$$

$\phi(t) \in \mathcal{C}([-\tau, 0], \mathbb{R}^n)$ is the initial *distributed state* where $\mathcal{C}([-\tau, 0], \mathbb{R}^n)$ denotes the set of functions $\phi : [-\tau, 0] \to \mathbb{R}^n$ which are continuous everywhere except for finite time instants $\bar{t}$ at which both $\phi(\bar{t}^-)$ and $\phi(\bar{t}^+)$ exist and $\phi(\bar{t}^-) = \phi(\bar{t})$. Denote by x_t the distributed state which is defined by

$$x_t(s) = x(t+s), \quad s \in [-\tau, 0].$$

In order to avoid Zeno solutions, time-regularization is assumed. Thus in any finite time interval, there are only finite reset time instants. Under this assumption, for any initial condition $\phi(t) \in \mathcal{C}([-\tau, 0], \mathbb{R}^n)$, the unique solution to the RCS (3.45) exists.

Theorem 3.10. *[7] If there exist positive matrices P and Q such that the functional defined by*

$$V(x_t) = x_t^T(0) P x_t(0) + \int_{-\tau}^{0} x_t^T(s) Q x_t(s) ds \tag{3.46}$$

satisfies the conditions

$$\frac{d}{dt} V(x_t) < 0, \quad x_t(0) \notin \ker C \tag{3.47}$$

$$\Delta V := V(x_{t^+}) - V(x_t) \leq 0, \quad x_t(0) \in \ker C, \tag{3.48}$$

then the RCS (3.45) is globally asymptotically stable.

Definition 3.7. *The RCS (3.45) is said to be quadratically stable if there exist positive matrices P and Q such that the conditions of Theorem 3.10 hold.*

Theorem 3.11. *[7] Suppose that the reset matrix R_ρ is Schur stable, i.e., $|\lambda(R_\rho)| < 1$. Then RCS (3.45) is quadratically stable if and only if there are matrices $Q = Q^T \in \mathbb{R}^{n\times n}$, $P_{11} = P_{11}^T \in \mathbb{R}^{n_p\times n_p}$, $P_{12} \in \mathbb{R}^{n_p\times(n_r-\rho)}$, $P_{22} = P_{22}^T \in \mathbb{R}^{(n_r-\rho)\times(n_r-\rho)}$, $P_{33} = P_{33}^T \in \mathbb{R}^{\rho\times\rho}$ and a vector $\beta \in \mathbb{R}^\rho$ such that*

$$P = \begin{bmatrix} P_{11} & P_{12} & C_p^T\beta^T \\ \star & P_{22} & 0_{(n_r-\rho)\times\rho} \\ \star & \star & P_{33} \end{bmatrix} > 0 \tag{3.49}$$

$$\begin{bmatrix} A^TP + PA + Q & PA_d \\ \star & -Q \end{bmatrix} < 0 \tag{3.50}$$

$$R_\rho^T P_{33} R_\rho - P_{33} \leq 0. \tag{3.51}$$

Proof: The *Lyapunov–Krasovskii functional* (3.46) can be alternatively expressed as

$$V(x_t) = x^T(t)Px(t) + \int_{-\tau}^{0} x^T(t+s)Qx(t+s)ds. \tag{3.52}$$

The time derivative of $V(x_t)$ can be easily obtained as

$$\begin{aligned} \frac{d}{dt}V(x_t) &= x^T(t)P\dot{x}(t) + \dot{x}^T(t)Px(t) + x^T(t)Qx(t) - x^T(t-\tau)Qx(t-\tau) \\ &= \begin{bmatrix} x(t) \\ x(t-\tau) \end{bmatrix}^T \begin{bmatrix} A^TP + PA + Q & PA_d \\ \star & -Q \end{bmatrix} \begin{bmatrix} x(t) \\ x(t-\tau) \end{bmatrix}. \end{aligned}$$

Thus (3.47) holds if and only if (3.50) is satisfied. In addition, we have

$$\Delta V = V(x_{t^+}) - V(x_t) = x^T(t)(R^TPR - P)x(t). \tag{3.53}$$

Thus (3.48) holds if and only if

$$(C^\perp)^T(R^TPR - P)C^\perp \leq 0 \tag{3.54}$$

where $C^\perp$ is a full-rank right annihilator of C which is the same as in (3.29). By following the same argument of the proof of Theorem 3.7, (3.48) holds if and only if P is subject to the structural constraint (3.49) and (3.51) holds.

Theorem 3.12. *[7] Suppose that $R_\rho = 0_{\rho\times\rho}$. Then the RCS (3.45) is quadratically stable if and only if there are matrices $P_\rho \in \mathbb{R}^{\rho\times\rho}$, $Q \in \mathbb{R}^{n\times n}$ and a vector $\beta \in \mathbb{R}^{\rho}$, such that for any $\omega \in \mathbb{R} \cup \{\infty\}$, there holds*

$$\begin{bmatrix} (j\omega I - A)^{-1}\mathcal{B} \\ I_{n+\rho} \end{bmatrix}^\star M \begin{bmatrix} (j\omega I - A)^{-1}\mathcal{B} \\ I_{n+\rho} \end{bmatrix} < 0 \tag{3.55}$$

where

$$\mathcal{B} = \left[\begin{array}{c|c} \begin{matrix} 0_{(n-\rho)\times\rho} \\ I_\rho \end{matrix} & A_d \end{array}\right], \quad \mathcal{C} = [\beta C_p \ \ 0_{\rho\times(n_r-\rho)} \ \ P_\rho] \tag{3.56}$$

$$M = \begin{bmatrix} Q & -\mathcal{C}^T & 0_{n\times n} \\ \star & -\epsilon I & 0_{\rho\times n} \\ \star & \star & -Q \end{bmatrix}.$$

Proof: It is easy to check that the conditions of Theorem 3.11 hold for the case $R_\rho = 0$ if and only if (3.50) holds for some P satisfying

$$\mathcal{C} = \mathcal{B}_1^T P \tag{3.57}$$

with $P_\rho \in \mathbb{R}^{\rho\times\rho}$, where $\mathcal{C}$ is defined in (3.56) and $\mathcal{B}_1 = [0_{\rho\times(n-\rho)} \ \ I_\rho]$. Thus quadratic stability amounts to the existence of P, $Q > 0$ such that (3.57) holds and

$$\begin{bmatrix} A^T P + PA + Q & PA_d \\ A_d^T P & -Q \end{bmatrix} < 0. \tag{3.58}$$

This is in turn equivalent to the existence of a P such that

$$\begin{bmatrix} A^T P + PA + Q & PA_d & P\mathcal{B}_1 - \mathcal{C}^T \\ \star & -Q & 0 \\ \star & \star & -\epsilon I \end{bmatrix} < 0, \quad \forall \epsilon > 0. \tag{3.59}$$

Interchanging the second and the third rows and columns, respectively, the above inequality is equivalent to

$$\begin{bmatrix} A^T P + PA + Q & P\mathcal{B}_1 - \mathcal{C}^T & PA_d \\ \star & -\epsilon I & \star \\ \star & 0 & -Q \end{bmatrix} < 0, \quad \forall \epsilon > 0, \tag{3.60}$$

i.e.,

$$M + \begin{bmatrix} A^T P + PA & P\mathcal{B} \\ \star & 0 \end{bmatrix} < 0, \quad \forall \epsilon > 0. \tag{3.61}$$

Thus the theorem is proven by using Lemma 3.3.

Remark 3.5. *If the conditions of Theorem 3.12 are satisfied, then the RCS (3.45) is quadratically stable for any reset matrix R_ρ such that*

$$R_\rho^T P_\rho R_\rho - P_\rho \leq 0.$$

3.4 Reset times-dependent stability

The quadratic stability analyzed in the previous sections implicitly assumes that the baseline system is asymptotically stable. However, an asymptotically stable RCS might be with an unstable baseline system [6, 8]. See also Remark 3.3 for an example of asymptotically stable RCSs with a marginally stable baseline system. Even for the case when the baseline system is asymptotically stable, a stable RCS is not necessarily quadratically stable [8]. The stability criteria introduced in the following do not assume stability of the baseline system *a priori*.

The basic idea of reset times-dependent stability is as follows. Consider a reset system given by

$$\begin{cases} \dot{x} = Ax, & x \notin \mathcal{J} \\ x^+ = Rx, & x \in \mathcal{J} \end{cases} \tag{3.62}$$

where $x \in \mathbb{R}^n$, $A, R \in \mathbb{R}^{n\times n}$, and $\mathcal{J}$ is the jump set. Assume that beating and Zenoness do not occur. If for any nonzero initial condition x_0, the solution to the reset system has infinitely many reset time instants, denoted by $\tau_i(x_0)$, $0 \leq i \leq \infty$, which can be arranged as

$$0 := \tau_0(x_0) < \tau_1(x_0) < \tau_2(x_0) < \cdots$$

where

$$\tau_i(x_0) - \tau_{i-1}(x_0) \geq \tau_m > 0,$$

then we can discretize the system at the after-reset time instants as

$$x(k+1) = Re^{A(\tau_k(x_0)-\tau_{k-1}(x_0))}x(k) \tag{3.63}$$

where $x(k) := x(\tau_k^+(x_0))$. By this way, it is possible to change the problem of stability analysis of reset systems to that of time-varying discrete-time systems. The benefit of this method is that it combines the baseline system and the reset action into a single system and the resulting stability criteria do not assume stability of the baseline system. The main difficulty of this method is how to determine whether any solution would intersect the jump set in finite time. For the case $\mathcal{J} = \ker C$ with C a row vector, this is the so-called continuous *Skolem–Pisot problem* which is described as follows: given a matrix $A \in \mathbb{R}^{n\times n}$, a row vector $C \in \mathbb{R}^{1\times n}$, and a initial state $x_0 \in \mathbb{R}^n$, determine whether there is a positive time instant t such that

$$Ce^{At}x_0 = 0.$$

The solvability of general Skolem–Pisot problem remains open [8]. In the following, we first consider the special continuous *Skolem–Pisot problem* where $A \in \mathbb{R}^{n\times n}$, $C \in \mathbb{R}^{1\times n}$, $x_0 \in \mathbb{R}^n$, and the matrix A is of complex dominant eigenvalues.

Lemma 3.4. *[9] For any constant μ, $Ce^{At}x_0 = 0$ if and only if $Ce^{(A+\mu I)t}x_0 = 0$.*

Proof: The proof follows by noting the facts that

$$Ce^{(A+\mu I)t}x_0 = e^{\mu t}Ce^{At}x_0$$

and $e^{\mu t} \neq 0$.

Lemma 3.4 tells that we can always assume that all of the eigenvalues of A have nonpositive real parts. For instance, if we choose

$$\mu = -\max\{\mathrm{Re}(\lambda) \mid \lambda \in \sigma(A)\},$$

then all of the eigenvalues of $A + \mu I$ will have zero or negative real parts.

In the following, J denotes the *Jordan normal form* of A. i.e.,

$$A = P^{-1}JP$$

with P the similarity matrix. Suppose that

$$J = \begin{bmatrix} J_1 & & & \\ & J_2 & & \\ & & \ddots & \\ & & & J_m \end{bmatrix}$$

where J_i is the ith *Jordan block* corresponding to the eigenvalue $\lambda_i \in \mathbb{C}$, i.e.,

$$J_i = \lambda_i I_{n_i} + M_i, \quad M_i = \begin{bmatrix} 0 & 1 & 0 & \cdots & 0 \\ 0 & 0 & 1 & \cdots & 0 \\ & & \ddots & \ddots & \vdots \\ & & & \ddots & 1 \\ & & & & 0 \end{bmatrix}_{n_i \times n_i} \tag{3.64}$$

with n_i the multiplicity of λ_i, $\sum_{i=1}^{m} n_i = n$. Thus we have

$$\begin{aligned} Ce^{At}x_0 &= CP^{-1}e^{Jt}Px_0 \\ &= CP^{-1}\begin{bmatrix} e^{J_1 t} & & & \\ & e^{J_2 t} & & \\ & & \ddots & \\ & & & e^{J_m t} \end{bmatrix} Px_0 \end{aligned}$$

Partition P and P^{-1} as

$$P = \begin{bmatrix} P_1 \\ P_2 \\ \vdots \\ P_m \end{bmatrix}, \quad P^{-1} = [Q_1 \;\; Q_2 \;\; \cdots \;\; Q_m]$$

where $P_i \in \mathbb{R}^{n_i \times n}$, $Q_i \in \mathbb{R}^{n \times n_i}$. Then

$$\begin{aligned} Ce^{At}x_0 &= \sum_{i-1}^{m} CQ_i e^{J_i t} P_i x_0 \\ &= \sum_{i-1}^{m} \left(CQ_i e^{M_i t} P_i x_0\right) e^{\lambda_i t} \\ &:= \sum_{i=1}^{m} p_i(t) e^{\lambda_i t} \end{aligned} \tag{3.65}$$

where $p_i(t)$s are polynomials with complex coefficients. The triple (A, C, x_0) is said to be *reduced* if for any i, the polynomial $p_i(t)$ is not identically zero [9]. Suppose that

$$\lambda_i = \alpha_i + j\beta_i$$

and

$$p_i(t) = p_{i,R}(t) + jp_{i,I}(t)$$

where $p_{i,R}(t)$ and $p_{i,I}(t)$ are polynomials with real coefficients. Note that $Ce^{At}x_0$ is a real function, we have

$$Ce^{At}x_0 = \sum_{i=1}^{m} e^{\alpha_i t}\left[p_{i,R}(t)\cos\beta_i t - p_{i,I}(t)\sin\beta_i t\right]. \tag{3.66}$$

Proposition 3.1. *[9] Suppose that (A, C, x_0) is reduced and none of the dominant eigenvalues of A is real. Then there is a positive time instant t such that $Ce^{At}x_0 = 0$.*

Proof: Only a sketched proof is provided here. More detailed proof can be found in [9]. By Lemma 3.4, we can assume that all of the dominant eigenvalues of A are with zero real parts, without loss of generality. By the assumption that none of the dominant eigenvalues of A is real, we can assume that all of the dominant eigenvalues are pure imaginary numbers. Suppose that there are $2k$ pure imaginary eigenvalues. Then by (3.66), $Ce^{At}x_0$ can be decomposed into two parts as

$$Ce^{At}x_0 = \sum_{i=1}^{2k}\left[p_{i,R}(t)\cos\beta_i t - p_{i,I}(t)\sin\beta_i t\right] + g(t) \tag{3.67}$$

where $g(t)$ is the term corresponding to the eigenvalues with negative real parts. Note that $g(t)$ tends to zero exponentially as $t \to \infty$ and the first part of the above formula oscillates since none of the polynomials $p_i(t)$ is identically zero. Actually, the first part is almost periodic [10, 9]. Thus there must exist a positive time instant t such that $Ce^{At}x_0 1 = 0$.

In the following, we consider the RCS (3.62) where the jump set is defined as

$$\mathcal{J} = \ker C \setminus \operatorname{Im} R,$$

and the after-reset set is denoted as $\mathcal{R}(\mathcal{J})$. Suppose that time-regularization is applied with the *time-regularization* constant $\tau_m > 0$. The detailed proof of the following proposition can be found in [8].

Proposition 3.2. *[8] Suppose that A does not have real dominant eigenvalues and $\mathcal{D}$ is a nonempty closed set of initial conditions satisfying $\mathcal{D} \subseteq \mathbb{R}^n \setminus \bar{\mathcal{R}}$, where*

$$\bar{\mathcal{R}} := \{x_0 \in \mathbb{R}^n \mid (A, C, x_0) \text{ is not reduced}\}. \tag{3.68}$$

Then, for any $x_0 \in \mathcal{D}$, the reset intervals are uniformly bounded, i.e., there is a positive constant τ_M such that

$$\tau_{k+1}(x_0) - \tau_k(x_0) \le \tau_M, \quad \forall k, \quad \forall x_0 \in \mathcal{D}. \tag{3.69}$$

Theorem 3.13. *[8] Suppose that none of the dominant eigenvalues of A is real and (A, C, x_0) is reduced for any nonzero initial state $x_0 \in \mathcal{R}(\mathcal{J})$. Then the RCS (3.62) with the initial state restricted to $\mathcal{R}(\mathcal{J})$ is asymptotically stable if the time-varying discrete-time system*

$$x_d(k+1) = Re^{A\Delta_k}x_d(k) \tag{3.70}$$

is uniformly asymptotically stable for any sequence $\{\Delta_k\}_{k\in\mathbb{Z}_{>0}}$ satisfying $\Delta_k \ge \tau_m$, $\forall k$.

Sketched Proof: Uniformly asymptotic stability of (3.70) implies that there are a constant $\gamma > 0$ and a constant λ, $0 \le \lambda < 1$, such that, for any initial state x_0, the after-reset state $x(\tau_k^+; x_0)$ satisfies

$$\|x(\tau_k^+; x_0)\| \le \gamma\lambda^k \|x_0\|. \tag{3.71}$$

On the other hand, for any k, there holds

$$x(t; x_0) = x(\tau_k^+; x_0) + \int_{\tau_k}^{t} Ax(s; x_0)ds. \tag{3.72}$$

Thus there is a positive constant $\alpha > 0$ such that

$$\|x(t; x_0)\| \le \|x(\tau_k^+; x_0)\| + \alpha \int_{\tau_k}^{t} \|x(s; x_0)\|ds, \quad t \in (\tau_k, \tau_{k+1}]. \tag{3.73}$$

By using Granwall inequality, there holds

$$\|x(t; x_0)\| \le \|x(\tau_k; x_0)\|e^{\alpha(t-\tau_k)}, \quad t \in (\tau_k, \tau_{k+1}]. \tag{3.74}$$

By Proposition 3.2, there is a positive number τ_M such that $\tau_{k+1} - \tau_k \le \tau_M$. Thus,

$$\|x(t;x_0)\| \le \gamma e^{\alpha\tau_M}\lambda^k \|x_0\|, \quad t \in (\tau_k, \tau_{k+1}]. \tag{3.75}$$

This implies that the RCS is asymptotically stable.

Theorem 3.14. *[8] Suppose that none of the dominant eigenvalues of A is real and (A, C, x_0) is reduced for any nonzero initial state $x_0 \in \mathcal{R}(\mathcal{J})$. If there is a sequence of positive definite matrices $\{P_k\}_{k\in\mathbb{Z}_{>0}}$ such that*

$$\eta I \le P_k \le \mu I, \quad \forall k$$

and

$$e^{A^T\Delta_k} R^T P_{k+1} R e^{A\Delta_k} - P_k \le -\varepsilon I \tag{3.76}$$

for some positive constants η, μ, and ε, and for any sequence of reset intervals $\{\Delta_k\}_{k\in\mathbb{Z}_{>0}}$ satisfying $\Delta_k \ge \tau_m$, then the RCS (3.62) with the initial state restricted to $\mathcal{R}(\mathcal{J})$ is asymptotically stable.

Remark 3.6. *If we can obtain the upper bound τ_M of the reset intervals* a priori, *then (3.76) is only required to hold for any sequence $\{\Delta_k\}_{k\in\mathbb{Z}_{>0}}$ satisfying $\tau_m \le \Delta_k \le \tau_M$. The same holds for the following corollary.*

Corollary 3.4. *Suppose that none of the dominant eigenvalues of A is real and (A, C, x_0) is reduced for any nonzero initial state $x_0 \in \mathcal{R}(\mathcal{J})$, and that the upper bound of the reset intervals is τ_M. If there exists a positive definite matrix P such that*

$$e^{A^T\Delta} R^T P R e^{A\Delta} - P \le 0, \quad \forall \Delta \in [\tau_m, \tau_M] \tag{3.77}$$

where τ_m is the time-regularization constant, then the RCS (3.62) with the initial state restricted to $\mathcal{R}(\mathcal{J})$ is asymptotically stable.

Remark 3.7. *If one of the following conditions holds, then under the conditions of Theorems 3.13 and 3.14 or Corollary 3.4, the RCS (3.62) without the initial state restriction $x_0 \in \mathcal{R}(\mathcal{J})$ is asymptotically stable:*

1. *The baseline system $\dot{x} = Ax$ is asymptotically stable;*
2. *The baseline system $\dot{x} = Ax$ is unstable, but the jump set $\mathcal{J}$ is reached from any initial state $x_0 \in \mathbb{R}^n \setminus \mathcal{J}$.*

Note that the set of the after-reset states $\mathcal{R}(\mathcal{J})$ has a linear subspace structure. If $x_0 \in \mathcal{R}(\mathcal{J})$ and δ is the smallest positive time instant t such that $Ce^{At}x_0 = 0$, then for any nonzero constant α, δ is also the smallest positive time instant t such that $Ce^{At}z_0 = 0$ where $z_0 = \alpha x_0$. Thus we can obtain the upper bound of the reset interval by searching all of the points x_0 on the unit sphere in $\mathcal{R}(\mathcal{J})$, i.e.,

$$x_0 \in \mathcal{S}^n \cap \mathcal{R}(\mathcal{J}) = \{x \in \mathcal{R}(\mathcal{J}) \mid \|x\| = 1\}.$$

Note that when Rank$(C) = 1$,

$$\dim \mathcal{J} = \dim \ker C = n - 1$$

and

$$\dim \mathcal{R}(\mathcal{J}) = n - 1 - \text{Rank}\,(R).$$

Suppose that Rank $(R) = n - 3$, then $\dim \mathcal{R}(\mathcal{J}) = 2$. In this case,

$$\dim \mathcal{S}^n \cap \mathcal{R}(\mathcal{J}) = 1.$$

This means that τ_M can be obtained by a single parameter search.

Another special case is that Rank $(R) = n - 2$ which leads to

$$\dim \mathcal{R}(\mathcal{J}) = 1.$$

This means that there is a nonzero vector $z \in \mathcal{R}(\mathcal{J})$ such that any element $x_0 = \mathcal{R}(\mathcal{J})$ can be represented by

$$x_0 = \mu z$$

for some number μ. Thus for any positive time instant t, there holds

$$Ce^{At}x_0 = 0$$

if and only if

$$Ce^{At}z = 0.$$

In this case, the reset time instants are equidistant. We have the following.

Corollary 3.5. *Suppose that* Rank $(R) = n - 2$ *and there is a positive number* δ *such that* $Ce^{A\delta}z = 0$ *where* z *is any vector in the after-reset set* $\mathcal{R}(\mathcal{J})$. *Without losing any generality, assume that* δ *is the smallest such number. Then the RCS (3.62) with the initial state restricted to* $\mathcal{R}(\mathcal{J})$ *is asymptotically stable if any only if* $Re^{A\delta}$ *is Schur stable, i.e.,*

$$|\lambda(Re^{A\delta})| < 1. \tag{3.78}$$

If in addition, one of the following conditions holds:

1. *The baseline system* $\dot{x} = Ax$ *is asymptotically stable;*
2. *The baseline system* $\dot{x} = Ax$ *is unstable, but the jump set* $\mathcal{J}$ *is reached from any initial state* $x_0 \in \mathbb{R}^n \setminus \mathcal{J}$,

then the RCS (3.62) is asymptotically stable.

For instance, consider a planar plant

$$\begin{cases} \dot{x}_p = A_p x_p + B_p u_p \\ y_p = C_p x_p \end{cases} \tag{3.79}$$

with $x_p = [x_{p1} \quad x_{p2}]^T \in \mathbb{R}^2$, $C_p = [1 \quad 0]$, controlled by the FORE

$$\begin{cases} \dot{x}_r = -bx_r + u_r, & x_p \notin \mathcal{J}_p \\ x_r^+ = 0, & x_p \in \mathcal{J}_p \end{cases} \tag{3.80}$$

through $u_p = x_r$, $u_r = -y_p$. The jump set is defined by

$$\mathcal{J}_p = \{x_p \in \mathbb{R}^2 \mid x_{p1} = 0, \quad x_{p2} \neq 0\}. \tag{3.81}$$

The closed-loop system is then given by

$$\begin{cases} \dot{x} = Ax, & x \notin \mathcal{J} \\ x^+ = Rx, & x \in \mathcal{J} \end{cases} \tag{3.82}$$

where

$$A = \begin{bmatrix} A_p & B_p \\ -C_p & -b \end{bmatrix}, \quad R = \begin{bmatrix} I_2 & 0 \\ 0 & 0 \end{bmatrix},$$

$$\mathcal{J} = \{x \in \mathbb{R}^3 \mid x_1 = 0, \quad x_2 \neq 0\}. \tag{3.83}$$

Obviously, we have

$$\mathcal{R}(\mathcal{J}) = \{x \in \mathbb{R}^3 \mid x_1 = 0, \quad x_2 \neq 0, \quad x_3 = 0\},$$

which is of dimension 1. Thus the reset time instants are equidistant. Take $z = [0 \quad 1 \quad 0]^T \in \mathcal{R}(\mathcal{J})$ and suppose that

$$e^{At} = [\phi_{ij}(t)]_{3\times 3}.$$

Then there holds

$$CRe^{At}z = \phi_{12}(t). \tag{3.84}$$

Suppose that $\delta > 0$ is the smallest positive number such that $\phi_{12}(t) = 0$. Then

$$\tau_i - \tau_{i-1} = \delta, \quad \forall i.$$

By Corollary 3.5, the RCS (3.82) with initial state being restricted in $\mathcal{R}(\mathcal{J})$ is asymptotically stable if and only if [11]

$$|\phi_{22}(\delta)| < 1.$$

Example 3.2. *Assume that* $n = 3$, $C = [1 \quad 0 \quad 0]$ *and*

$$A = \begin{bmatrix} -1 & 2 & 1 \\ -2 & -1 & 0 \\ 0 & 0 & -2 \end{bmatrix}, \quad R = \begin{bmatrix} 1 & 0 & 0 \\ 0 & 1 & 0 \\ 0 & 0 & r \end{bmatrix}$$

where r *is a constant. Suppose that*

$$\mathcal{J} = \ker C \setminus \{0\}.$$

It is easy to check that the set of after-reset states is

$$\mathcal{R}(\mathcal{J}) = \left\{ \begin{bmatrix} 0 \\ x_2 \\ rx_3 \end{bmatrix} \middle| \, x_2 \neq 0 \text{ or } x_3 \neq 0 \right\}.$$

If $r \neq 0$, we have

$$\mathcal{R}(\mathcal{J}) = \left\{ \begin{bmatrix} 0 \\ x_2 \\ x_3 \end{bmatrix} \middle| \, x_2 \neq 0 \text{ or } x_3 \neq 0 \right\}.$$

Thus we can numerically calculate τ_M by searching for all of the initial states x_0 satisfying

$$x_0 \in \mathcal{S}^3 \cap \mathcal{R}(\mathcal{J}) = \left\{ \begin{bmatrix} 0 \\ \cos\phi \\ \sin\phi \end{bmatrix} \middle| \, 0 \leq \phi \leq 2\pi \right\}.$$

On the other hand, note that $\lambda(A) = \{-1 \pm 2i, \ -2\}$. By using Lemma 3.4, we can replace A by

$$\bar{A} := A - \max \operatorname{Re} \lambda(A) = A + I. \tag{3.85}$$

After some calculations, we have

$$\bar{y}(t,\phi) := Ce^{\bar{A}t}x_0 = \cos\phi \sin 2t + \sin\phi \left[\frac{1}{5}\cos 2t + \frac{2}{5}\sin 2t - \frac{1}{5}e^{-t} \right].$$

Suppose that the time-regularization constant is $\tau_m > 0$ and $\tau_M(\phi)$ is the smallest positive number t such that $t \geq \tau_m$ and $\bar{y}(t,\phi) = 0$. Then

$$\tau_M = \max_{\phi \in [0,2\pi]} \tau_M(\phi).$$

The curve $\tau_M(\phi)$, $\phi \in [0, 2\pi]$, for the case with $\tau_m = 0.25$ is depicted in Figure 3.6 from which we have $\tau_M \approx 2.304$. Take $r = -0.5$,

$$Q = \begin{bmatrix} 1 & 0 & 0 \\ 0 & 1 & 0 \\ 0 & 0 & 0.2 \end{bmatrix}$$

and solve the Lyapunov function

$$e^{A^T \tau_m} R^T P R e^{A\tau_m} - P = -Q$$

for P by using Matlab tool box. We have

$$P = \begin{bmatrix} 2.5415 & 0 & 0.2753 \\ 0 & 2.5415 & 0.0405 \\ 0.2753 & 0.0405 & 0.2725 \end{bmatrix}.$$

We use this P to check the conditions of Corollary 3.4. Define

$$\bar{\lambda}_M(t) = \lambda_{\max}(e^{A^T t} RPRe^{At} - P).$$

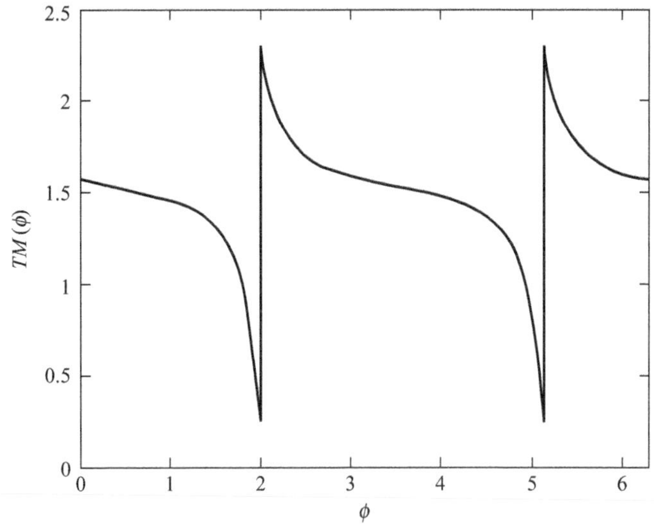

Figure 3.6 Reset interval for different initial state from the set of after-reset states

For different r, the curves of $\bar{\lambda}_M(t)$, $t \in [\tau_m, \tau_M]$, are depicted in Figure 3.7. Simple search shows that

$$\bar{\lambda}_M(t) < 0, \quad \forall t \in [\tau_m, \tau_M], \forall r \in [-1, 1].$$

Note that the baseline system is asymptotically stable, thus this reset system is asymptotically stable. The state response of the reset system for $x(0) = [2 \quad -1 \quad 6]^T$ and $r = -0.5$ is given in Figure 3.8.

For the case with $r = 0$, we have

$$\mathcal{R}(\mathcal{J}) = \left\{ \begin{bmatrix} 0 \\ x_2 \\ 0 \end{bmatrix} \middle| x_2 \neq 0 \right\}.$$

In this case, $\mathcal{R}(\mathcal{J})$ is a subset of the after-reset state set for the case with $r \neq 0$, thus it has been naturally included in the analysis for $r \neq 0$. In the following, we consider this case separately. Take

$$x_0 = [0 \;\; 1 \;\; 0]^T \in \mathcal{R}(\mathcal{J}),$$

then we have

$$\bar{y}(t, x_0) := Ce^{\bar{A}t}x_0 = \sin 2t$$

where $\bar{A}$ is defined in (3.85). Thus the reset interval is given by

$$\delta = \min\{t \mid t \geq \tau_m, \quad \sin 2t = 0\}.$$

For instance, if $\tau_m = 0.25$, then we have $\delta = \pi/2$. Thus the reset system is asymptotically stable since the baseline system is stable and $|\lambda(Re^{A\delta})| < 1$.

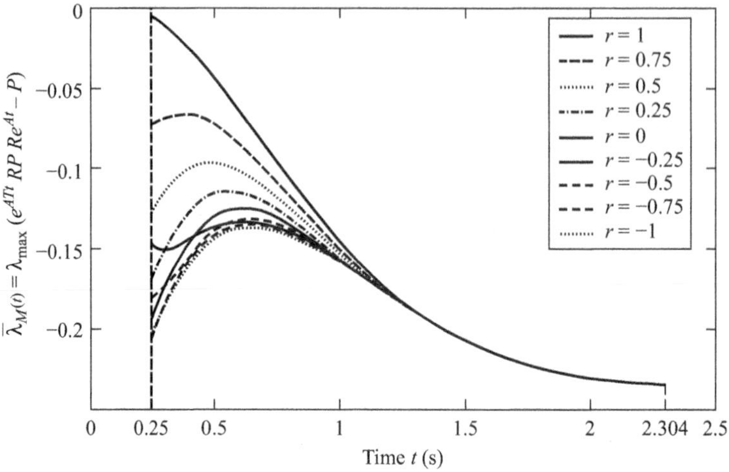

Figure 3.7 Max eigenvalue of $e^{A^T t}RPRe^{At} - P$ for different r

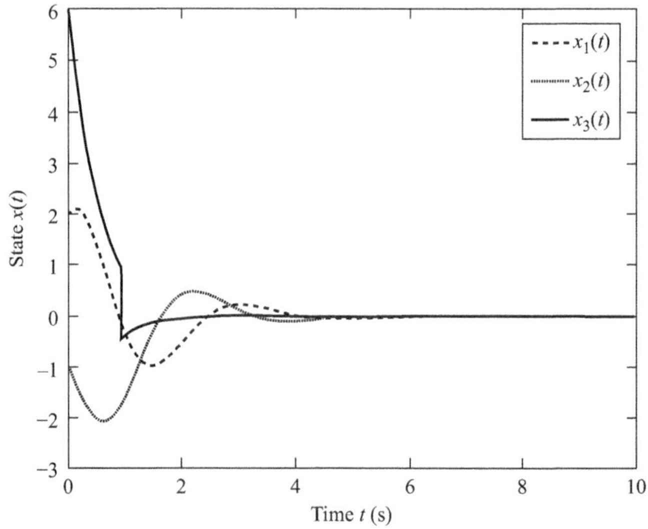

Figure 3.8 State response for $x(0) = [2 \quad -1 \quad 6]^T$ and $r = -0.5$

Remark 3.8. *The main advantage of the reset times-dependent stability criteria in this section is that they do not assume stability of the baseline system. Thus they are applicable to a subset of stable RCSs with unstable baseline systems. In addition, they are also applicable to a subset of RCSs with stable baseline systems which are not quadratically stable. See Reference 8 for the detail.*

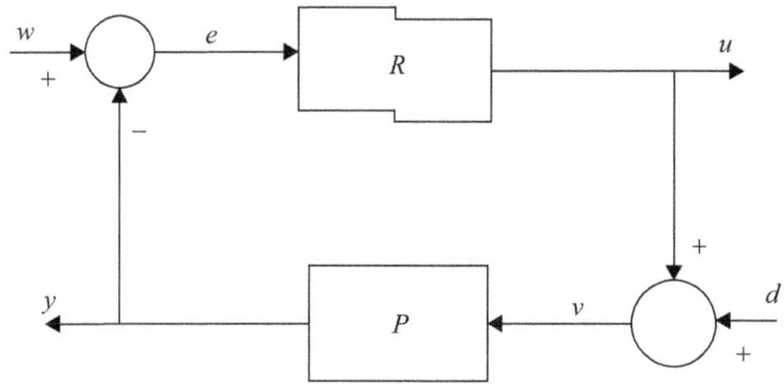

Figure 3.9 A reset control system

3.5 Passivity of RCSs

Consider an RCS depicted in Figure 3.9 where the plant is described by

$$\begin{cases} \dot{x}_p = f(x_p, v) \\ y = g(x_p, v) \end{cases} \tag{3.86}$$

and the reset controller R is given by

$$\begin{cases} \dot{x}_r = A_r x_r + B_r e, & e \neq 0 \\ x_r^+ = R_r x_r, & e = 0. \\ u = C_r x_r + D_r e. \end{cases} \tag{3.87}$$

The plant and the controller are interconnected via

$$e = w - y, \quad v = u + d$$

where w and d are reference and external disturbance, respectively. The reset controller is assumed to be time-regularized to avoid Zeno solutions. According to the Passivity Theorem (Theorem 3.2), the RCS is finite $\mathcal{L}_2$-gain stable from (w, d) to (u, y) if both the plant and the reset controller satisfy certain passivity properties. The following results give the conditions the reset controller should satisfy to possess certain passivity property.

Theorem 3.15. *[5] The reset controller (3.87) with $R_r = 0_{n_r \times n_r}$ is passive, input strictly passive, output strictly passive, or very strictly passive if the baseline controller*

$$\begin{cases} \dot{x}_r = A_r x_r + B_r e \\ u = C_r x_r + D_r e. \end{cases} \tag{3.88}$$

is passive, input strictly passive, output strictly passive, or very strictly passive, respectively.

Proof: Only a sketched proof for very strictly passivity is provided here. For any $T \geq 0$, suppose that $\{\tau_i\}_{1\leq i\leq k_T}$ is the set of reset time instants over $[0, T]$ with

$$0 := \tau_0 < \tau_1 < \tau_2 < \cdots < \tau_{k_L} < \tau_{k_L+1} := T.$$

The condition $R_r = 0_{n_r\times n_r}$ implies that at any reset time instant, there holds $x_r(\tau_i^+) = 0$ and $e^T(\tau_i^+)u(\tau_i^+) = 0$. Thus, by the assumption that the baseline controller is very strictly passive, there exist constants $\delta > 0$ and $\varepsilon > 0$ such that

$$\int_{\tau_i}^{\tau_{i+1}} e^T(s)u(s)ds \geq \delta \int_{\tau_i}^{\tau_{i+1}} e^T(s)e(s)ds + \varepsilon \int_{\tau_i}^{\tau_{i+1}} u^T(s)u(s)ds, \quad \forall e \in \mathcal{L}_{2,e}, \tag{3.89}$$

holds for any $0 \leq i \leq k_T$. Summing up these inequalities from $i = 0$ to $i = k_T$ leads to

$$\int_0^T e^T(s)u(s)ds \geq \delta \int_0^T e^T(s)e(s)ds + \varepsilon \int_0^T u^T(s)u(s)ds. \tag{3.90}$$

Thus the reset controller is very strictly passive.

By using Theorems 3.3 and 3.4, we have the following:

Corollary 3.6.
1. *The reset controller (3.87) with $R_r = 0_{n_r\times n_r}$ is passive if the baseline transfer function is positive real;*
2. *The reset controller (3.87) with $R_r = 0_{n_r\times n_r}$ is very strictly passive if the baseline transfer function is strictly positive real and there are matrices $P = P^T > 0$, L and W such that*

$$\begin{bmatrix} -A_r^TP - PA_r + L^TL & L^TW \\ W^TL & D + D^T \end{bmatrix} > 0. \tag{3.91}$$

By using Theorem 3.5, we have the following result.

Corollary 3.7. *Consider reset controller (3.87) with $R_r = 0_{n_r\times n_r}$ and $\mathrm{Re}\,[\lambda(A_r)] < 0$. Denote*

$$h_r(s) = C_r(sI - A_r)B_r + D_r.$$

Then, the following hold:

1. *The reset controller (3.87) is passive if*

   $$\mathrm{Re}\,[h_r(j\omega)] \geq 0, \quad \forall\omega \in [-\infty, +\infty];$$

2. *The reset controller (3.87) is input strictly passive if there is a constant $\delta > 0$ such that*

   $$\mathrm{Re}\,[h_r(j\omega)] \geq \delta > 0, \quad \forall\omega \in [-\infty, +\infty];$$

3. *The reset controller (3.87) is output strictly passive if there is constant $\varepsilon > 0$ such that*

   $$\mathrm{Re}\,[h_r(j\omega)] \geq \varepsilon|h_r(j\omega)|^2 > 0, \quad \forall\omega \in [-\infty, +\infty].$$

For reset controller (3.87) with general reset matrix R_r, we have the following.

Theorem 3.16. *[5] The reset controller (3.87) is passive, input strictly passive, output strictly passive, or very strictly passive if the baseline controller (3.88) is dissipative with respect to supply rates $w_p(e,u)$, $w_i(e,u)$, $w_o(e,u)$, or $w_v(e,u)$, respectively, and with a storage function $V(x_r)$ satisfying*

$$V(R_r x_r) \leq V(x_r), \quad \forall x \in \mathbb{R}^{n_r}, \tag{3.92}$$

where $w_p(e,u)$, $w_i(e,u)$, $w_o(e,u)$, and $w_v(e,u)$ are defined as:

1. $w_p(e,u) = e^T u$;
2. $w_i(e,u) = e^T u - \delta e^T e$, $\delta > 0$;
3. $w_o(e,u) = e^T u - \varepsilon u^T u$, $\varepsilon > 0$; *and*
4. $w_v(e,u) = e^T u - \delta e^T e - \varepsilon u^T u$, $\delta > 0$, $\varepsilon > 0$,

Sketched Proof: For any $T \geq 0$, suppose that $\{\tau_i\}_{1 \leq i \leq k_T}$ is the set of reset time instants over $[0, T]$ with

$$0 := \tau_0 < \tau_1 < \tau_2 < \cdots < \tau_{k_L} < \tau_{k_L+1} := T.$$

If the baseline controller is dissipative with respective to the supply rate

$$w(e,u) = e^T u - \delta e^T e - \varepsilon u^T u$$

with $\delta \geq 0$, $\varepsilon > 0$, then we have

$$\begin{aligned} V(x_r(\tau_{i+1})) &\leq V(x_r(\tau_i^+)) + \int_{\tau_i}^{\tau_{i+1}} w(e(s), u(s))ds \\ &\leq V(R_r x_r(\tau_i)) + \int_{\tau_i}^{\tau_{i+1}} w(e(s), u(s))ds \\ &\leq V(x_r(\tau_i)) + \int_{\tau_i}^{\tau_{i+1}} w(e(s), u(s))ds, \quad \forall i = 0, 1, \ldots, k_T. \end{aligned}$$

Thus there holds

$$0 \leq V(x_r(T)) \leq V(x_r(0)) + \int_0^T w(e(s), u(s))ds$$

which in turn implies that

$$\int_0^T e^T(s)u(s)ds \geq \beta + \delta \int_0^T e^T(s)e(s)ds + \varepsilon \int_0^T u^T(s)u(s)ds$$

with $\beta := -V(x_r(0))$.

Example 3.3. *Consider a FORE*

$$\begin{cases} \dot{x}_r = -ax_r + be, & e \neq 0 \\ x_r^+ = rx_r, & e = 0 \\ u = cx_r + de, \end{cases} \tag{3.93}$$

where $x_r \in \mathbb{R}^1$ and a is a positive constant. Take

$$V(x_r) = \frac{\lambda}{2}x_r^2$$

as the storage function candidate, then

$$\begin{aligned}\dot{V} &= -\lambda a x_r^2 + \lambda b e x_r \\ &= -\frac{\lambda a}{c^2}(u - de)^2 + \frac{\lambda b}{c}e(u - de) \\ &= \frac{\lambda(bc + 2ad)}{c^2}eu - \frac{\lambda d(bc + ad)}{c^2}e^2 - \frac{\lambda a}{c^2}u^2.\end{aligned}$$

By taking $\lambda = \frac{c^2}{bc+2ad}$, it is obtained that

$$\dot{V} = eu - \frac{d(bc + ad)}{bc + 2ad}e^2 - \frac{a}{bc + 2ad}u^2$$

Noting that

$$V(rx_r) - V(x_r) = \frac{\lambda}{2}(r^2 - 1)x_r^2 \leq 0, \quad \forall |r| \leq 1,$$

we have the following results:

1. *The FORE is passive and output strictly passive for any $|r| \leq 1$ if*

   $$d = 0, \quad bc > 0.$$

2. *The FORE is input strictly passive and very strictly passive for any $|r| \leq 1$ if*

   $$\begin{cases} bc + 2ad > 0 \\ d(bc + ad) > 0. \end{cases}$$

   Or equivalently,

   $$ad > 0, \quad bc + ad > 0.$$

By using Theorem 3.4, we have the following result.

Corollary 3.8. *Consider the reset controller (3.87) with*

$$R_r = \begin{bmatrix} I_{n_r-\rho} & 0 \\ 0 & R_\rho \end{bmatrix}$$

where R_ρ is a $\rho \times \rho$ matrix which is assumed to be Schur stable, i.e., $|\lambda(R_\rho)| < 1$. Then the reset controller is very strictly passive if there are matrices $P_1 = P_1^T > 0$, $P_2 = P_2^T > 0$, L, and W such that

$$\begin{bmatrix} -A_r^T P - PA_r + L^T L & L^T W \\ W^T L & D + D^T \end{bmatrix} > 0. \tag{3.94}$$

$$R_\rho P_2 R_\rho - P_2 \leq 0, \tag{3.95}$$

where

$$P = \begin{bmatrix} P_1 & 0 \\ 0 & P_2 \end{bmatrix}. \tag{3.96}$$

Proof: By Corollary 3.6, the baseline controller is dissipative with respect to supply rate $w_v(e,u) = e^T u - \delta e^T e - \varepsilon u^T u$ and the storage function $V(x_r) = x_r^T P x_r$ (See Reference 2). It is easy to check that, for this storage function, $V(R_r x_r) \leq V(x_r)$ if and only if P is of the form (3.96) and (3.95) holds. Thus by Theorem 3.16, the reset controller is very strictly passive.

By using the Passivity Theorem (Theorem 3.2) and the results above, we can obtain different sufficient conditions for the RCS depicted in Figure 3.9 to be finite $\mathcal{L}_2$-gain stable.

Theorem 3.17. *[5] The RCS depicted in Figure 3.9 with the reset controller given by (3.87) is finite $\mathcal{L}_2$-gain stable if one of the following conditions is satisfied:*

1. *The plant is input strictly passive and the baseline controller (3.88) is dissipative with respect to the supply rate $w_i(e,u)$ and a storage function $V(x_r)$ satisfying (3.92);*
2. *The plant is output strictly passive and the baseline controller (3.88) is dissipative with respect to the supply rate $w_o(e,u)$ and a storage function $V(x_r)$ satisfying (3.92);*
3. *The plant is passive and the baseline controller (3.88) is dissipative with respect to the supply rate $w_v(e,u)$ and a storage function $V(x_r)$ satisfying (3.92);*
4. *The plant is very strictly passive and the baseline controller (3.88) is dissipative with respect to the supply rate $w_p(e,u)$ and a storage function $V(x_r)$ satisfying (3.92),*

where w_i, w_o, w_v , and w_p are defined in Theorem 3.16.

Notes

The results on quadratic stability of RCSs included in this chapter are mainly based on Reference 6 where the H_β condition was proved. The results regarding stability of RCSs with time-delays come from Reference 7. Less conservative results for RCSs with time-delays can be found in Reference 12 which have not been included in this book. The reset times-dependent stability results are mainly based on Reference 8. We also refer the reader to Reference 9 for the solvability of Skolem–Pisot problem which is crucial in establishing the reset times-dependent results in Reference 8. The results on passivity of reset controllers and finite $\mathcal{L}_2$-gain stability can be found in Reference 5. The basic concepts and results of dissipative systems come from References 2 and 1. Also, we refer the reader to Reference 13 for more detailed discussions on well-posedness and stability of RCSs with zero reset matrices.

References

[1] A. van der Schaft. L_2-gain and passivity in nonlinear control. In *Communications and Control Engineering*, Springer-Verlag London Ltd., Berlin, Heidelberg, New York, 2000.

[2] B. Brogliato, R. Lozano, B. Maschke, and O. Egeland. Dissipative systems analysis and control. In *Communications and Control Engineering*, Springer Science & Business Media, 2006.

[3] M. Vidyasagar. Nonlinear systems analysis. In *Classics in Applied Mathematics*, vol. 42. Society for Industrial and Applied Mathematics, Philadelphia, Pennsylvania, 2002.

[4] D. Nešic, L. Zaccarian, and A.R. Teel. Stability properties of reset systems. *Automatica*, 44(8):2019–2026, 2008.

[5] J. Carrasco, A. Baños, and A. van der Schaft. A passivity-based approach to reset control systems stability. *Systems & Control Letters*, 59(1):18–24, 2010.

[6] O. Beker, C.V. Hollot, Y. Chait, and H. Han. Fundamental properties of reset control systems. *Automatica*, 40(6):905–915, 2004.

[7] A. Baños and A. Barreiro. Delay-independent stability of reset systems. *IEEE Transactions on Automatic Control*, 54(2):341–346, 2009.

[8] A. Baños, J. Carrasco, and A. Barreiro. Reset times-dependent stability of reset control systems. *IEEE Transactions on Automatic Control*, 56(1):217–223, 2011.

[9] P.C. Bell, J. Delvenne, R.M. Jungers, and V.D. Blondel. The continuous Skolem–Pisot problem. *Theoretical Computer Science*, 411(40):3625–3634, 2010.

[10] H. Bohr. *Almost periodic functions*, Chelsea Publishing Company, 1951.

[11] Q. Chen, C.V. Hollot, Y. Chait, and O. Beker. On reset control systems with second-order plants. In *Proceedings of the American Control Conference*, June 28–30, 2000, vol. 1, pp. 205–209. IEEE, Chicago, IL, 2000.

[12] A. Barreiro and A. Baños. Delay-dependent stability of reset systems. *Automatica*, 46(1):216–221, 2010.

[13] A. Baños and A. Barreiro. Reset control systems. In *Advances in Industrial Control*, Springer-Verlag London Limited, 2011.

Chapter 4
Robust stability of reset control systems

One of the major difficulties in stability analysis of reset systems comes from uncertain reset time instants due to uncertain parameters. For instance, consider the case when a reset action is triggered whenever the tracking error crosses zero. If there are uncertain (constant or time-varying) parameters in the output matrix of a plant, then the time instants of zero-crossings also become uncertain. In practice, a well-designed *reset control system* (RCS) should be expected to be robust with respect to such uncertainty. This chapter is devoted to robust stability of RCSs with uncertainties in output matrices. The focus will be on single-output systems. Two kinds of uncertainties will be discussed. The first kind is bounded uncertain time-varying parameters and the second kind is uncertain constant parameters. Robust stability of RCSs with time-delay will also be discussed.

4.1 Definitions and assumptions

Consider the uncertain RCS consisting of the plant

$$\Sigma_p(\Delta): \quad \begin{cases} \dot{x}_p = A_p x_p + B_p u \\ y = (C_p + \Delta_c) x_p \end{cases} \tag{4.1}$$

with $x \in \mathbb{R}^{n_p}$, $u \in \mathbb{R}^m$, $y \in \mathbb{R}^1$, and the reset controller

$$\Sigma_R: \quad \begin{cases} \dot{x}_r = A_r x_r + B_r y, & y \neq 0 \\ x_r^+ = R_r x_r, & y = 0 \\ u = F_r x_r + G_r y \end{cases} \tag{4.2}$$

with $x_r \in \mathbb{R}^{n_r}$. In (4.1), $\Delta_c \in \Omega_c \subset \mathbb{R}^{1 \times n_p}$ is the uncertainty in the output matrix where Ω_c is a bounded subset of $\mathbb{R}^{1 \times n_p}$ called *the uncertainty set*. The nominal output matrix C_p is assumed to be of full rank, i.e.,

$$\text{Rank}\,(C_p) = 1.$$

Assume that the size of the uncertainty is smaller than the nominal value in the sense of

$$\|\Delta_c\| < \|C_p\|$$

where $\|\cdot\|$ represents the *Euclidean norm*. Note that we can always assume that

$$C_p = [\,1 \quad 0 \quad \cdots \quad 0\,] \tag{4.3}$$

without losing any generality. In fact, if this is not the case, we can always find an $n_p \times n_p$ orthogonal matrix $\tilde{S}$ with $\tilde{S}^T\tilde{S} = I$ such that

$$C_p\tilde{S}^{-1} = [\,\|C_p\| \quad 0 \quad \cdots \quad 0\,].$$

Then $z_p = Sx_p$ with $S = \|C_p\|\tilde{S}$ transforms (4.1) to

$$\begin{cases} \dot{z}_p = SA_pS^{-1}z_p + SB_pu, \\ y = (C_p + \Delta_c)S^{-1}z_p \end{cases} \tag{4.4}$$

with

$$C_pS^{-1} = [\,1 \quad 0 \quad \cdots \quad 0\,]$$

and

$$\|\Delta_cS^{-1}\| = \|C_p\|^{-1}\|\Delta_c\| < 1 = \|C_pS^{-1}\|.$$

Thus, without losing any generality, we assume that C_p is already of the form (4.3). The reset matrix R_r is assumed to be of the block-diagonal form

$$R_r = \begin{bmatrix} I_{n_r-\rho} & \\ & R_\rho \end{bmatrix}$$

where R_ρ is a $\rho \times \rho$ matrix with $\rho \leq n_r$ and $I_{n_r-\rho}$ is the identity matrix with dimension $n_r - \rho$. This structure indicates that part of the controller states is reset. In the early development of RCSs, the reset matrix R_ρ is assumed to be zero. Here, we only assume that R_ρ is Schur stable in the sense that all of its eigenvalues lie within the open unit disc, i.e.,

$$|\lambda(R_\rho)| < 1.$$

Denote

$$\Delta = \begin{bmatrix} \Delta_c & 0_{1\times n_r} \end{bmatrix}, \quad C = \begin{bmatrix} C_p & 0_{1\times n_r} \end{bmatrix}, \quad C_\Delta = C + \Delta.$$

Combining (4.1) with (4.2) gives

$$\begin{cases} \dot{x} = A_\Delta x, & C_\Delta x \neq 0 \\ x^+ = Rx, & C_\Delta x = 0, \end{cases} \tag{4.5}$$

where $x = \begin{bmatrix} x_p \\ x_r \end{bmatrix} \in \mathbb{R}^n$ with $n = n_p + n_r$ and

$$A_\Delta = A + H\Delta$$

with

$$A = \begin{bmatrix} A_p + B_pG_rC_p & B_pF_r \\ B_rC_p & A_r \end{bmatrix}, \quad H = \begin{bmatrix} B_pG_r \\ B_r \end{bmatrix}, \quad R = \begin{bmatrix} I & \\ & R_r \end{bmatrix}.$$

Remark 4.1. *In order to avoid Zeno behaviors and ensure well-posedness, the reset rate is assumed to be limited by time-regularization [1, 2], i.e., the reset action is triggered only when the time spent since the latest reset action is greater than a pre-specified positive number $\tau_{\min}$. Thus there are only a finite number of reset actions during any compact time interval.*

Remark 4.2. *Instead of system (4.1), one can consider more general uncertain linear plant*

$$\begin{cases} \dot{x}_p = (A_p + \Delta_a)x_p + (B_p + \Delta_b)u, \\ y = (C_p + \Delta_c)x_p \end{cases} \tag{4.6}$$

with $\Delta_A \in \Omega_A \subset \mathbb{R}^{n_p \times n_p}$, $\Delta_B \in \Omega_B \subset \mathbb{R}^{n_p \times m}$, and $\Delta_c \in \Omega_c \subset \mathbb{R}^{1 \times n_p}$. However, among the uncertainties Δ_a, Δ_b, and Δ_c, we only focus on the output uncertainty Δ_c because the output uncertainty Δ_c not only affects the structure matrix of the closed-loop system but also makes the reset time instants uncertain. So long as we solve the problem of robust stability with respect to Δ_c, then Δ_a and Δ_b can be treated by using existing tools of robustness analysis, for instance, the method reported in Reference 3. Thus we just assume that $\Delta_a = 0$ and $\Delta_b = 0$ for simplicity.

Remark 4.3. *Since C_p is of the form (4.3), the assumption $\|\Delta_c\| < \|C_p\|$ implies that*

$$|\delta_1| < \|C_p\| = 1, \quad \forall \Delta_c \in \Omega_c \tag{4.7}$$

where δ_1 is the first component of Δ_c. Note that the condition (4.7) is less conservative than $\|\Delta_c\| < \|C_p\|$ and we shall see that all the results in this chapter still hold if the assumption $\|\Delta_c\| < \|C_p\|$ is replaced by (4.7).

Two kinds of uncertainties will be considered. The first one is time-varying uncertainty Δ_c. The second one is constant uncertainty Δ_c. For the former, we consider quadratic stability in the following sense.

Definition 4.1. *RCS (4.5) is said to be quadratically stable with respect to Ω_c if there is a positive definite matrix P such that:*

1. *For any $\Delta_c \in \Omega_c$, there holds*

$$A_\Delta^T P + PA_\Delta < 0. \tag{4.8}$$

2. *For any reset time instant t_k, there holds*

$$x^T(t_k^+)Px(t_k^+) - x^T(t_k)Px(t_k) \leq 0. \tag{4.9}$$

From the definition above, one sees that quadratic stability requires that both the baseline system and the reset action be dissipative with respect to a common quadratic energy function which is independent of the uncertainty.

For the constant uncertainty case, the notion of quadratic stability becomes conservative. We consider *affine quadratic stability* in the following sense instead which assures asymptotic stability for any constant uncertainty.

Definition 4.2. *Suppose that*

$$\Delta_c = [\delta_1 \ \cdots \ \delta_{n_p}].$$

Then the RCS (4.5) is said to be affine quadratically stable if there is a parameter-dependent matrix P_Δ of the form

$$P_\Delta = P_0 + \sum_{i=1}^{n_p} \delta_i P_i \tag{4.10}$$

such that

1. *For any $\Delta_c \in \Omega_c$, there holds*

$$A_\Delta^T P_\Delta + P_\Delta A_\Delta < 0. \tag{4.11}$$

2. *For any reset time instant t_k, there holds*

$$x^T(t_k^+) P_\Delta x(t_k^+) - x^T(t_k) P_\Delta x(t_k) \le 0. \tag{4.12}$$

4.2 Quadratic stability

For simplicity, we denote $\Delta_c = [\delta_1, \gamma]$ where δ_1 is a scalar and

$$\gamma = [\delta_2, \ldots, \delta_{n_p}]$$

is a row vector with dimension $n_p - 1$. It is not difficult to show that the reset system (4.5) is quadratically stable if and only if there is a positive definite matrix P such that

$$A_\Delta^T P + P A_\Delta < 0, \quad \forall \Delta_c \in \Omega_c, \tag{4.13}$$

$$x^T(R^T P R - P)x \le 0, \quad \forall x \in \bigcup_{\Delta_c \in \Omega_c} \ker(C_\Delta). \tag{4.14}$$

Suppose that such a matrix P exists and $\Delta_c(t) \in \Omega_c$ is any function. Take

$$V(x) = x^T P x$$

as a Lyapunov function candidate. Then on one hand, if t is not a reset time instant, i.e., $C_{\Delta(t)} x(t) \neq 0$, then

$$\dot{V}(x) = x^T (A_{\Delta(t)}^T P + P A_{\Delta(t)}) x < 0.$$

On the other hand, if t is a reset time instant, then $C_{\Delta(t)} x(t) = 0$ and thus

$$x(t) \in \ker(C_{\Delta(t)}) \subseteq \bigcup_{\Delta_c \in \Omega_c} \ker(C_\Delta).$$

So, there holds

$$V(x(t^+)) - V(x(t) = x^T(t)(R^T P R - P)x(t) \le 0.$$

This means that the RCS is asymptotically stable for any time-varying uncertainty $\Delta_c(t) \in \Omega_c$.

4.2.1 RCSs with low-dimensional plants ($n_p \leq 2$)

4.2.1.1 Scalar plant ($n_p = 1$)

In this case, the closed-loop RCS is of the form (4.5) with

$$C = [1\ 0\ \cdots\ 0], \quad \Delta_c = [\delta_1\ 0\ \cdots\ 0], \quad R = \begin{bmatrix} 1 & 0 \\ 0 & R_r \end{bmatrix}. \tag{4.15}$$

Note that the reset time instances are not affected by the uncertainty in this case since $(C + \Delta_c)x = 0$ if and only if $x_1 = 0$, where $x = [x_1, x_2^T]^T$ with $x_1 \in \mathbb{R}^1$ and $x_2 \in \mathbb{R}^{n_p+n_r-1}$. Thus (4.14) holds if and only if

$$x_2^T(R_r^T P_{22} R_r - P_{22})x_2 \leq 0, \quad \forall x_2 \in \mathbb{R}^{n_p+n_r-1}$$

where

$$P = \begin{bmatrix} p_{11} & p_{12} \\ p_{12}^T & P_{22} \end{bmatrix}. \tag{4.16}$$

This is in turn equivalent to $R_r^T P_{22} R_r - P_{22} \leq 0$. Thus we have the following.

Proposition 4.1. *Consider RCS (4.5) whose parameters are given in (4.15). Suppose that the uncertainty set Ω_c is given by*

$$\delta_1 \in [\underline{\delta},\ \bar{\delta}] := \Omega_\delta.$$

Then this RCS is quadratically stable if and only if there is a positive definite matrix P which is of the form (4.16) such that

$$A_{\delta_1}^T P + P A_{\delta_1} < 0, \quad \delta_1 = \underline{\delta} \text{ and } \bar{\delta} \tag{4.17}$$

$$R_r^T P_{22} R_r - P_{22} \leq 0, \tag{4.18}$$

where A_{δ_1} represents A_Δ for this case.

4.2.1.2 Two-dimensional plant ($n_p = 2$)

When $n_p = 2$, the uncertainty output matrix of the RCS (4.5) is given by

$$C_\Delta = [1 + \delta_1\ \ \delta_2\ \ 0\ \ \cdots\ \ 0].$$

In this case, $C_\Delta x = 0$ if and only if

$$(1 + \delta_1)x_1 + \delta_2 x_2 = 0.$$

Define

$$k_{\max} := \sup_{(\delta_1,\delta_2)\in\Omega_c} \left[-\frac{\delta_2}{1+\delta_1} \right],$$

$$k_{\min} := \inf_{(\delta_1,\delta_2)\in\Omega_c} \left[-\frac{\delta_2}{1+\delta_1} \right].$$

Define an $n \times n$ symmetric matrix $\tilde{Q}_\Omega$ as

$$\tilde{Q}_\Omega = \left[\begin{array}{c|c} Q_\Omega & 0_{2\times(n-2)} \\ \hline 0_{(n-2)\times 2} & 0_{(n-2)\times(n-2)} \end{array}\right]$$

where

$$Q_\Omega = \begin{bmatrix} 1 & -\frac{1}{2}(k_{\max} + k_{\min}) \\ -\frac{1}{2}(k_{\max} + k_{\min}) & k_{\max}k_{\min} \end{bmatrix}.$$

It is trivial to check that

$$\bigcup_{\Delta_c \in \Omega_c} \ker(C_\Delta) \subseteq \{x \in \mathbb{R}^n \mid x^T \tilde{Q}_\Omega x \leq 0\}.$$

If the uncertainty set Ω_c is a compact and convex subset, then the above inclusion becomes an equality, i.e.,

$$\bigcup_{\Delta_c \in \Omega_c} \ker(C_\Delta) = \{x \in \mathbb{R}^n \mid x^T \tilde{Q}_\Omega x \leq 0\}. \tag{4.19}$$

By $\mathcal{S}$-procedure [4], (4.14) holds if and only if there is a nonnegative number $\mu \geq 0$ such that

$$R^T PR - P - \mu\tilde{Q}_\Omega \leq 0,$$

which is equivalent to

$$\begin{bmatrix} \mu Q_\Omega & P_{12}(I_2 - R_r) \\ (I_2 - R_r^T)P_{12}^T & P_{22} - R_r^T P_{22} R_r \end{bmatrix} \geq 0. \tag{4.20}$$

where

$$P = \begin{bmatrix} P_{11} & P_{12} \\ P_{12}^T & P_{22} \end{bmatrix}. \tag{4.21}$$

By the definition of Q_Ω, the (1, 1)-element of Q_Ω is positive and

$$\det(Q_\Omega) = -\frac{1}{4}(k_{\max} - k_{\min})^2 \leq 0.$$

- **Case 1 (the trivial case):** $k_{\max} = k_{\min}$.
  In this case, $-\frac{\delta_2}{1+\delta_1}$ is a constant, denoted by k. That is, the uncertainty does not affect the reset time instants since

  $$\bigcup_{\Delta_c \in \Omega_c} \ker(C_\Delta) = \{x \in \mathbb{R}^n \mid x_1 = kx_2\}.$$

  If Ω_c contains the origin, then we have $\delta_2 = 0$, i.e., $k = 0$. Thus it can be treated in the same way as in the scalar plant case.

- **Case 2:** $k_{\max} \neq k_{\min}$.
  In this case, in order for (4.20) to hold, μ must be zero in which case (4.20) is equivalent to

  $$\begin{cases} P_{12}(I - R_r) = 0 \\ R_r^T P_{22} R_r - P_{22} \leq 0. \end{cases}$$

  Denote

  $$r_1 = \text{Rank}\,(I - R_r).$$

  By Lemma 3.1, $P_{12}(I - R_r) = 0$ is equivalent to the existence of a $2 \times (n - 2 - r_1)$ matrix Γ such that

  $$P_{12} = \Gamma \left[(I - R_r^T)^{\perp} \right]^T.$$

  We have the following.

Proposition 4.2. *Consider RCS (4.5) with $n_p = 2$. Suppose that the uncertainty set Ω_c is a compact convex subset of $\mathbb{R}^2$ containing at least one element (δ_1, δ_2) with $\delta_2 \neq 0$. Then RCS (4.5) is quadratically stable if and only if there are a 2×2 symmetric matrix P_{11}, a $(n-2) \times (n-2)$ symmetric matrix P_{22}, and a $2 \times (n - 2 - r_1)$ matrix Γ such that*

$$P = \begin{bmatrix} P_{11} & \Gamma \left[(I - R_r^T)^{\perp} \right]^T \\ \left[(I - R_r^T)^{\perp} \right] \Gamma^T & P_{22} \end{bmatrix} > 0,$$

$$A_\Delta^T P + P A_\Delta < 0, \quad \forall \Delta_c \in \text{Ver}(\Omega_c),$$

$$R_r^T P_{22} R_r - P_{22} \leq 0$$

where $\text{Ver}(\Omega_c)$ *represents the set of vertices of* Ω_c.

4.2.2 High-dimensional cases

For the case with $n_p \geq 3$, the structure of $\bigcup_{\Delta_c \in \Omega_c} \ker(C_\Delta)$ is much more complicated and it is in general difficult to find a $\tilde{Q}_\Omega$ such that (4.19) holds. Thus the technique proposed in Section 4.2.1 does not apply to this case. The following results for the case with $n_p \geq 3$ are obtained by using the technique of *parameter-dependent full-rank annihilators* of matrices.

Proposition 4.3. *Suppose that Ω_c is a convex polyhedron containing the origin and at least one $\Delta_c = [\delta_1, \gamma]$ with $\gamma \neq 0$. Then the RCS (4.5) is quadratically stable with respect to Ω_c if and only if there exists a positive definite block-diagonal matrix*

$$P = \begin{bmatrix} P_{11} & \\ & P_{22} \end{bmatrix} \tag{4.22}$$

where the dimensions of P_{11} *and* P_{22} *are* $(n_p+n_r-\rho)\times(n_p+n_r-\rho)$ *and* $\rho\times\rho$, *respectively, such that*

$$A_\Delta^T P + PA_\Delta < 0, \quad \forall \Delta_c \in \mathrm{Ver}(\Omega_c), \tag{4.23}$$

$$R_\rho^T P_{22} R_\rho - P_{22} \le 0. \tag{4.24}$$

Here, $\mathrm{Ver}(\Omega_c)$ *represents the set of vertices of* Ω_c.

Proof: We first prove that (4.14) is equivalent to the condition that there is a positive definite block-diagonal matrix P of the form (4.22) such that (4.24) holds. The key to prove this claim is to construct a full-rank right annihilator of C_Δ, denoted by $C_\Delta^\perp$, which is continuously dependent on Δ_c. $C_\Delta^\perp$ is nothing but a $(n_p+n_r)\times(n_p+n_r-1)$ matrix satisfying

$$C_\Delta C_\Delta^\perp = 0$$

and

$$\mathrm{Rank}\,(C_\Delta^\perp) = n_p + n_r - 1$$

for any $\Delta_c \in \Omega_c$. In other words, $C_\Delta^\perp$ is a full-rank matrix satisfying

$$\ker\,(C_\Delta) = \mathrm{Im}\,(C_\Delta^\perp)$$

for any $\Delta_c \in \Omega_c$ with $\ker\,(\cdot)$ and $\mathrm{Im}\,(\cdot)$ representing the kernel and image spaces of matrices, respectively. Thus, if such a $C_\Delta^\perp$ is constructed, (4.14) is equivalent to

$$(C_\Delta^\perp)^T (R^T PR - P) C_\Delta^\perp \le 0, \quad \forall \Delta_c \in \Omega_c. \tag{4.25}$$

By the assumption that $\|\Delta_c\| < \|C_p\| = 1$, there holds that $1+\delta_1 \neq 0$. Based on this, it is easy to check that one of the full-rank right annihilators of C_Δ is given by

$$C_\Delta^\perp = \begin{bmatrix} J_\Delta & \\ & I_\rho \end{bmatrix} \tag{4.26}$$

with I_ρ a $\rho\times\rho$ identity matrix and J_Δ a $(n_p+n_r-\rho)\times(n_p+n_r-\rho-1)$ block-diagonal matrix defined as

$$J_\Delta = \left[\begin{array}{c|c} \begin{bmatrix} \gamma \\ -(1+\delta_1) I_{n_p-1} \end{bmatrix} & \\ \hline & I_{n_r-\rho} \end{array}\right]. \tag{4.27}$$

The fact $1+\delta_1 \neq 0$ assures that J_Δ is indeed of full column rank. Repartition R and P as

$$R = \begin{bmatrix} I_{n_p+n_r-\rho} & \\ & R_\rho \end{bmatrix}, \quad P = \begin{bmatrix} P_{11} & P_{12} \\ P_{12}^T & P_{22} \end{bmatrix},$$

then straightforward calculation shows that (4.25) is equivalent to

$$\left[\begin{array}{c|c} 0 & J_\Delta^T P_{12}(R_\rho - I) \\ \hline \star & R_\rho^T P_{22} R_\rho - P_{22} \end{array}\right] \le 0, \quad \forall \Delta_c \in \Omega_c$$

which is in turn equivalent to

$$R_\rho^T P_{22} R_\rho - P_{22} \le 0, \tag{4.28}$$

$$J_\Delta^T P_{12}(R_\rho - I) = 0, \quad \forall \Delta_c \in \Omega_c. \tag{4.29}$$

Since R_ρ is Schur stable, $R_\rho - I$ is nonsingular and thus (4.29) is equivalent to

$$J_\Delta^T P_{12} = 0, \quad \forall \Delta_c \in \Omega_c. \tag{4.30}$$

Partition P_{12} according to the partition of J_Δ as

$$P_{12} = \begin{bmatrix} \Gamma_1 \\ \Gamma_2 \\ \Gamma_3 \end{bmatrix},$$

where Γ_1, Γ_2, and Γ_3 are, respectively, $1 \times \rho$, $(n_p - 1) \times \rho$, and $(n_r - \rho) \times \rho$ matrices. Then (4.30) is equivalent to $\Gamma_3 = 0$ and

$$\gamma^T \Gamma_1 - (1 + \delta_1)\Gamma_2 = 0. \tag{4.31}$$

By the assumption that Ω_c contains the origin, we can choose $\Delta_c = [\delta_1, \gamma] = 0$. Thus we get $\Gamma_2 = 0$ and (4.30) is equivalent to

$$\gamma^T \Gamma_1 = 0, \quad \Gamma_2 = 0, \quad \Gamma_3 = 0.$$

Since Ω_c contains at least one element $\Delta_c = [\delta, \gamma]$ with $\gamma \neq 0$, then $\gamma^T \Gamma_1 = 0$ for any $\Delta_c \in \Omega_c$ if and only if $\Gamma_1 = 0$. Based on the observation above, (4.30) is equivalent to $P_{12} = 0$, i.e., P is block-diagonal. Finally, according to the convexity of Ω_c and the affine dependence of A_Δ on Δ, it is easy to see that P satisfies (4.13) if and only if it satisfies (4.23). Thus the proof is completed.

Remark 4.4. *In the proof of Proposition 4.3, the full-rank right annihilator of C_Δ is not unique. But the result does not depend on the choice of $C_\Delta^\perp$. In fact, suppose that Q_1 and Q_2 are two different choices of $C_\Delta^\perp$, then*

$$\begin{cases} C_\Delta Q_i = 0 \\ \text{Rank}\,(Q_i) = n_p + n_r - 1, \end{cases} \quad i = 1, 2. \tag{4.32}$$

Thus there holds that

$$\text{Im}\,(Q_1) = \text{Im}\,(Q_2) = \ker\,(C_\Delta)$$

which implies that there is a nonsingular square matrix Y such that $Q_1 = Q_2 Y$. So, the replacement of $C_\Delta^\perp$ in (4.25) by Q_1 or Q_2 leads to two equivalent matrix inequalities.

Remark 4.5. *Note that the technique of full-rank right annihilators of matrices has been used to establish the H_β-condition for the uncertainty-free case with $R_\rho = 0$. See Chapter 3. In this chapter, the choice of the uncertain parameter-dependent annihilator (4.26) is the key step to prove Proposition 4.3 and subsequent results.*

Remark 4.6. *The convex polyhedron Ω_c is defined as the convexity closure of a finite set $\{\Delta_1, \ldots, \Delta_N\} \subset \mathbb{R}^{1\times n_p}$, i.e.,*

$$\Omega_c = \overline{co}\,\{\Delta_1, \ldots, \Delta_N\}.$$

Thus it includes the special case that some components of C_p do not have uncertainty, i.e., some components of Δ_c are zero. Refer to the illustrative example in Section 4.5.

In the above propositions, it is required that the uncertainty set Ω_c contains at least one $\Delta_c = [\delta_1\ \gamma]$ with $\gamma \neq 0$. This requirement rules out the special case that all the elements of Ω_c have the form

$$\Delta_c = [\delta_1\ 0\ \cdots\ 0],$$

i.e., the uncertainty set Ω_c degenerates to

$$\delta_1 \in [\underline{\delta},\ \bar{\delta}] := \Omega_\delta.$$

Though this special case is trivial, we shall see in the following Proposition 4.4 that we cannot arrive at $P_{12} = 0$ any more.

Proposition 4.4. *Suppose that all the elements of Ω_c are of the form $\Delta_c = [\delta_1, \gamma]$ with $\gamma = 0$ and $\delta_1 \in \Omega_\delta := [\underline{\delta}, \bar{\delta}]$. Then the necessary and sufficient condition for the RCS (4.5) to be quadratically stable with respect to Ω_δ is that there exist symmetric matrices P_{11}, P_{22}, and a column vector μ such that*

$$P := \begin{bmatrix} P_{11} & e_1\mu^T \\ \star & P_{22} \end{bmatrix} > 0 \tag{4.33}$$

and

$$A_{\delta_1}^T P + PA_{\delta_1} < 0, \quad \delta_1 = \underline{\delta} \text{ and } \bar{\delta} \tag{4.34}$$

$$R_\rho^T P_{22} R_\rho - P_{22} \leq 0, \tag{4.35}$$

where e_1 is the $(n_p + n_r - \rho)$-dimensional column vector whose first element is 1 and the others are zero, A_{δ_1} stands for A_Δ for this case.

Proof: By following the same argument of Proposition 4.3, we only need to prove that (4.14) implies

$$P_{12} = e_1\mu^T.$$

Note that in this case, the output of the system becomes

$$y = (C_p + \Delta_c)x = (1 + \delta_1)x_1.$$

One easily sees that $(1+\delta_1)x_1 = 0$ if and only if $x_1 = 0$ provided that $|\delta_1| < 1$. This means that the uncertainty δ_1 has no direct impact on the reset time instants. Accordingly, inequality (4.25) becomes

$$(C^\perp)^T(R^TPR - P)C^\perp \leq 0 \tag{4.36}$$

whose left-hand side is independent of the uncertain parameter δ_1. By taking

$$C^{\perp} = [\, 0 \quad I_{n_p+n_r-1} \,]^T$$

and following the same argument of Proposition 4.3, the above inequality is equivalent to (4.28) and $J^T P_{12} = 0$ with $J = [\, 0 \quad I_{n_p+n_r-\rho-1} \,]^T$. Thus P_{12} has a representation

$$P_{12} = (J^T)^{\perp} \mu^T$$

for some $\rho \times 1$ vector μ and a full-rank right annihilator $(J^T)^{\perp}$ of J^T. Note that one of the choices for $(J^T)^{\perp}$ is $(J^T)^{\perp} = e_1$. So, $P_{12} = e_1 \mu^T$.

Remark 4.7. *Note that if $R_\rho = 0$, the condition (4.35) naturally holds. If in addition the system is uncertainty-free, i.e., $\underline{\delta} = \bar{\delta}$, the conditions (4.33) and (4.34) in Proposition 4.4 are equivalent to the H_β-condition (Theorem 3.10).*

If (4.6) is the system under consideration, we can obtain the following result by following the same argument.

Proposition 4.5. *Consider the RCS consisting of the plant (4.6) and the controller (4.2):*

1. *If Ω_c contains the origin and at least one $\Delta_c = [\delta, \gamma]$ with $\gamma \neq 0$, then the RCS is quadratically stable if and only if the baseline system is quadratically stable with respect to Ω_c with a block-diagonal Lyapunov matrix P as in (4.22) satisfying (4.24).*
2. *If all the elements of Ω_c are of the form $\Delta_c = [\delta_1, \gamma]$ with $\gamma = 0$ and $\delta_1 \in \Omega_\delta := [\underline{\delta}, \bar{\delta}]$, then the RCS is quadratically stable if and only if the baseline system is quadratically stable with respect to Ω_δ with a Lyapunov matrix P as in (4.33) satisfying (4.35).*

4.3 Affine quadratic stability

In the following, we turn to consider the conditions under which the RCS is asymptotically stable for any constant $\Delta_c \in \Omega_c$. For simplicity, we assume that Ω_c is a hyper-rectangle whose center is at $\Delta_{center} \in \Omega_c$. One easily sees that if for any $\Delta_c \in \Omega$, there is a positive definite matrix P_Δ depending on Δ_c such that

$$A_\Delta^T P_\Delta + P_\Delta A_\Delta < 0, \quad \forall \Delta_c \in \Omega_c, \tag{4.37}$$

$$(C_\Delta^{\perp})^T (R^T P_\Delta R - P_\Delta) C_\Delta^{\perp} \leq 0, \quad \forall \Delta_c \in \Omega_c \tag{4.38}$$

with $C_\Delta^{\perp}$ given in (4.26), then the RCS is asymptotically stable for any $\Delta_c \in \Omega_c$. Note that A_Δ is affine with respect to $\Delta_c = [\delta_1 \ \cdots \ \delta_{n_p}]$. More precisely,

$$A_\Delta = A + H\Delta = A + \sum_{i=1}^{n_p} \delta_i H E_i^T \tag{4.39}$$

where E_i represents the ith column of the identity matrix $I_{n_p+n_r}$. We aim to find the conditions under which there is a P_Δ of the affine form

$$P_\Delta = P_0 + \sum_{i=1}^{n_p} \delta_i P_i \tag{4.40}$$

such that (4.37) and (4.38) hold.

In the following, we first investigate the structural constraints that should be imposed on such a P_Δ. Suppose that

$$P_\Delta = \begin{bmatrix} P_{11}(\Delta) & P_{12}(\Delta) \\ \star & P_{22}(\Delta) \end{bmatrix} \tag{4.41}$$

where P_{11}, P_{12}, and P_{22} are matrices of dimensions $(n_p + n_r - \rho) \times (n_p + n_r - \rho)$, $(n_p + n_r - \rho) \times \rho$, and $\rho \times \rho$, respectively. Partition each P_i accordingly as

$$P_i = \begin{bmatrix} \Gamma_{11}^i & \Gamma_{12}^i \\ \star & \Gamma_{22}^i \end{bmatrix}, \quad i = 0, 1, \ldots, n_p. \tag{4.42}$$

According to the analysis in Section 4.2, the inequality (4.38) is equivalent to

$$R_\rho^T P_{22}(\Delta) R_\rho - P_{22}(\Delta) \le 0, \quad \forall \Delta_c \in \Omega_c \tag{4.43}$$

$$J_\Delta^T P_{12}(\Delta) = 0, \quad \forall \Delta_c \in \Omega_c, \tag{4.44}$$

where J_Δ is given in (4.27). It is easy to check that one of the full-rank right annihilators of J_Δ^T is given by

$$(J_\Delta^T)^\perp = \begin{bmatrix} 1+\delta_1 & \gamma & 0_{1\times(n_r-\rho)} \end{bmatrix}^T.$$

By (4.44), $P_{12}(\Delta)$ can be represented as

$$P_{12}(\Delta) = (J_\Delta^T)^\perp \mu^T,$$

where μ is a column vector with dimension ρ. Denoted by e_i $(i = 1, 2, \ldots, n_p)$ the ith column of the identity matrix $I_{n_p+n_r-\rho}$. With this notation, $P_{12}(\Delta)$ can also be represented as

$$P_{12}(\Delta) = e_1 \mu^T + \sum_{i=1}^{n_p} \delta_i e_i \mu^T. \tag{4.45}$$

On the other hand, in view of (4.10), (4.41), and (4.42),

$$P_{12}(\Delta) = \Gamma_{12}^0 + \sum_{i=1}^{n_p} \delta_i \Gamma_{12}^i. \tag{4.46}$$

Comparing (4.46) with (4.45) leads to $\Gamma_{12}^0 = e_1 \mu^T$ and

$$\Gamma_{12}^i = e_i \mu^T, \quad i = 1, 2, \ldots, n_p$$

which constitute the structural constraints on P_Δ.

Another problem is that the left-hand side of (4.37) is not a convex combination of $\Delta_c \in \text{Ver}(\Omega_c)$. However, if it is *multi-convex* in the sense that it is convex with respect to each component of Δ_c, then (4.37) is equivalent to a finite set of LMIs which is indicated in the following lemma.

Lemma 4.1. *[5] Consider a scalar quadratic function of* $\Delta_c = [\delta_1, \ldots, \delta_{n_p}]$:

$$f(\Delta_c) = \alpha_0 + \sum_{i=1}^{n_p} \alpha_i \delta_i + \sum_{i<j} \beta_{ij} \delta_i \delta_j + \sum_{i=1}^{n_p} \eta_i \delta_i^2$$

and assume that $f(\cdot)$ *is multi-convex, i.e.,*

$$2\eta_i = \frac{\partial^2 f(\Delta_c)}{\partial \delta_i^2} \geq 0, \quad i = 1, \ldots, n_p.$$

Then $f(\Delta_c)$ *is negative in the hyper-rectangle* Ω_c *if and only if*

$$f(\Delta_c) < 0, \quad \forall \Delta_c \in \text{Ver}(\Omega_c).$$

Now we are ready to state the result as follows.

Proposition 4.6. *Consider the RCS (4.5) with a hyper-rectangle uncertainty set* Ω_c. *Suppose that* A_Δ *is stable at the center* $\Delta_c = \Delta_{center}$ *of* Ω_c. *If there are* $2(n_p + 1)$ *symmetric matrices* Γ_{jj}^i, $i = 0, 1, \ldots, n_p$, $j = 1, 2$, *and a* $\rho \times 1$ *vector* μ *such that*

$$\begin{aligned} P_\Delta &= \begin{bmatrix} \Gamma_{11}^0 & e_1 \mu^T \\ \star & \Gamma_{22}^0 \end{bmatrix} + \sum_{i=1}^{n_p} \delta_i \begin{bmatrix} \Gamma_{11}^i & e_i \mu^T \\ \star & \Gamma_{22}^i \end{bmatrix} \\ &:= P_0 + \sum_{i=1}^{n_p} \delta_i P_i \end{aligned} \tag{4.47}$$

satisfies

$$A_\Delta^T P_\Delta + P_\Delta A_\Delta < 0, \quad \forall \Delta_c \in \text{Ver}(\Omega_c), \tag{4.48}$$

$$E_i H^T P_i + P_i H E_i^T \geq 0, \quad i = 1, 2, \ldots, n_p, \tag{4.49}$$

$$R_\rho^T P_{22}(\Delta) R_\rho - P_{22}(\Delta) \leq 0, \quad \forall \Delta_c \in \text{Ver}(\Omega_c), \tag{4.50}$$

with

$$P_{22}(\Delta) = \Gamma_{22}^0 + \sum_{i=1}^{n_p} \delta_i \Gamma_{22}^i,$$

then the RCS (4.5) is affinely quadratically stable.

Proof: In order to prove this proposition, we only need to show that $P_\Delta > 0$ for any $\Delta_c \in \Omega_c$ and that (4.37) and (4.38) hold.

First, by (4.49) and following the same argument of Theorem 3.2 in Reference 5, for any x, the function

$$f(x,\Delta_c) := x^T(A_\Delta^T P_\Delta + P_\Delta A_\Delta)x$$

is multi-convex with respect to Δ_c. In fact, we have

$$\frac{\partial^2 f(x,\Delta_c)}{\partial \delta_i^2} = 2x^T(E_i H^T P_i + P_i H E_i^T)x \geq 0.$$

By (4.48) and Lemma 4.1, $f(x,\Delta_c) < 0$ for any $x \neq 0$ and $\Delta_c \in \Omega_c$. Thus (4.37) follows.

Second, since (4.37) holds and A_Δ is stable at $\Delta_{center} \in \Omega_c$, we have P_Δ is positive definite at Δ_{center}. Note that (4.37) implies that P_Δ is nonsingular at any $\Delta_c \in \Omega$, we have $P_\Delta > 0$ for any $\Delta_c \in \Omega$.

Lastly, by the analysis at the beginning of this section, (4.38) is equivalent to (4.43) and (4.44). Note that the left-hand side of (4.50) is affine with respect to Δ_c and that Ω_c is convex, (4.50) implies (4.43). Finally, by the structure of P_Δ in (4.47), $P_{12}(\Delta)$ can be represented as $P_{12}(\Delta) = (J_\Delta^T)^\perp \mu^T$. Thus, for any $\Delta_c \in \Omega_c$, $J_\Delta^T P_{12}(\Delta) = 0$, i.e., (4.44) holds.

Remark 4.8. *Consider that the conditions of Proposition 4.6 do not hold for the whole uncertainty set Ω_c. If, however, Ω_c is covered by a collection of small hyper-rectangles as $\Omega_c \subseteq \bigcup_j \Omega_j$ and every hyper-rectangle Ω_j meets the conditions of Proposition 4.6, then the RCS is asymptotically stable for any $\Delta_c \in \Omega_c$. Note that in this case, Ω_c is not required to be convex.*

4.4 Robust stability of RCS with time-delay

Consider an RCS consisting of a single-output plant with output uncertainty

$$\Sigma_p: \quad \begin{cases} \dot{x}_p = A_p x_p + B_p u_p \\ y_p = (C_p + \Delta_c) x_p \end{cases} \tag{4.51}$$

with $x_p \in \mathbb{R}^{n_p}$, $u_p \in \mathbb{R}^{m_p}$, $y_p \in \mathbb{R}^1$, and a reset controller with temporal regularization [1]

$$\Sigma_r: \quad \begin{cases} \dot{x}_r = A_r x_r + B_r u_r, & u_r \neq 0 \text{ or } \delta < \delta_m \\ x_r^+ = R_r x_r, & u_r = 0 \text{ and } \delta \geq \delta_m \\ y_r = C_r x_r + D_r u_r \end{cases} \tag{4.52}$$

with $x_r \in \mathbb{R}^{n_r}$, $u_r \in \mathbb{R}^1$, $y_r \in \mathbb{R}^{m_p}$. The scalar variable δ represents the time spent since the latest reset and is governed by the dynamics

$$\begin{cases} \dot{\delta} = 1, \ \delta(0) = 0, & u_r \neq 0 \text{ or } \delta < \delta_m \\ \delta^+ = 0, & u_r = 0 \text{ and } \delta \geq \delta_m, \end{cases} \tag{4.53}$$

where δ_m is a positive constant used to avoid Zeno solutions by limiting the reset rate. In (4.51), $\Delta_c \in \Omega_c \subset \mathbb{R}^{1\times n_p}$ is the uncertainty in the output matrix where Ω_c is a bounded convex subset of $\mathbb{R}^{1\times n_p}$. The reset matrix R_r in (4.52) is assumed to be of the form

$$R_r = \begin{bmatrix} I_{n_r-\rho} & \\ & R_\rho \end{bmatrix} \tag{4.54}$$

where $I_{n_r-\rho}$ is the $(n_r-\rho)\times(n_r-\rho)$ identity matrix and R_ρ is a $\rho\times\rho$ matrix ($\rho \le n_r$). The plant (4.51) and the reset controller (4.52) are interconnected through

$$\begin{cases} u_p(t) = y_r(t-\tau) \\ u_r = -y_p \end{cases} \tag{4.55}$$

where τ is the controller-to-actuator delay. Combining (4.51), (4.52) , and (4.55) leads to a closed-loop system of the form

$$\begin{cases} \dot{x} = A_{\Delta_c}x + E_{\Delta_c}x(t-\tau), & C_{\Delta_c}x \neq 0 \text{ or } \delta < \delta_m \\ x^+ = Rx, & C_{\Delta_c}x = 0 \text{ and } \delta \ge \delta_m \end{cases} \tag{4.56}$$

with $x = (x_p^T \ \ x_r^T)^T \in R^n$, $n = n_p + n_r$, $C_{\Delta_c} = [C_p + \Delta_c \ \ 0]$, and

$$A_{\Delta_c} = \begin{bmatrix} A_p & 0 \\ -B_r(C_p+\Delta_c) & A_r \end{bmatrix}, \tag{4.57}$$

$$E_{\Delta_c} = \begin{bmatrix} -B_pD_r(C_p+\Delta_c) & B_pC_r \\ 0 & 0 \end{bmatrix}, \tag{4.58}$$

$$R = \begin{bmatrix} I_{n_p} & \\ & R_r \end{bmatrix}. \tag{4.59}$$

Denote by $\mathcal{PC}([-\tau,0],\mathbb{R}^n)$ the set of piecewise continuous functions $\phi:[-\tau,0]\to\mathbb{R}^n$ with a finite number of step discontinuities [6], endowed with the norm

$$\|\phi\|_{\mathcal{PC}} := \max_{s\in[-\tau,0]} \|\phi(s)\|, \quad \forall\phi \in \mathcal{PC}([-\tau,0],\mathbb{R}^n),$$

where $\|\cdot\|$ is the Euclidean norm. For any $\phi \in \mathcal{PC}([-\tau,0],\mathbb{R}^n)$, define ϕ^+ as

$$\phi^+(\theta) = \begin{cases} \phi(\theta), & \theta \in [-\tau,0) \\ R\phi(0), & \theta = 0 \end{cases} \tag{4.60}$$

where R is the reset matrix defined in (4.59). Obviously, $\phi \in \mathcal{PC}([-\tau,0],\mathbb{R}^n)$ implies $\phi^+ \in \mathcal{PC}([-\tau,0],\mathbb{R}^n)$. We call the mapping $\phi \mapsto \phi^+$ the *reset mapping*. For any function $\psi:[-\tau,\infty)\to\mathbb{R}^n$ and $t\in[0,\infty)$, define $\psi_t:[-\tau,0]\to\mathbb{R}^n$ as

$$\psi_t(\theta) = \psi(t+\theta), \quad \theta \in [-\tau,0].$$

For any $\Delta_c \in \Omega_c$, denote by $x(t,\phi_0,\Delta_c)$ (or $x(t)$ for short) the solution to RCS (4.56) with the initial state $\phi_0 \in \mathcal{PC}([-\tau,0],\mathbb{R}^n)$ and initial time instant $t_0 = 0$, which is supposed to be continuous from the left at any reset time instant and continuous within intervals defined by successive reset time instants. We always assume that such a solution exists and only consider the admissible uncertainty $\Delta_c(t) \in \Omega_c$ which does not violate the existence of solutions. Thus for any $\phi_0 \in \mathcal{PC}([-\tau, 0], \mathbb{R}^n)$, there holds

$$x_t(\cdot, \phi_0, \Delta_c) \in \mathcal{PC}([-\tau, 0], \mathbb{R}^n), \quad \forall t \geq 0.$$

In this chapter, we are concerned with robust stability of RCS (4.56) in the sense of being asymptotically stable for any admissible time-varying uncertainty $\Delta_c(t) \in \Omega_c$.

Similar to the case without time-delays, we make the following assumptions.

Assumption 4.1. *The output matrix C_p is of the form*

$$C_p = [1 \;\; 0 \;\; \cdots \;\; 0].$$

Assumption 4.2. *The uncertainty set Ω_c contains the origin and satisfies*

$$|\delta_1| < 1, \quad \forall \Delta_c = [\delta_1 \;\; \gamma] \in \Omega_c,$$

where δ_1 is the first component of Δ_c and $\gamma = [\delta_2 \;\; \cdots \;\; \delta_{n_p}]$.

The proof of the following *generalized Lyapunov–Krasovskii Theorem* is similar to that in References 7, 8, and 9.

Proposition 4.7. *Suppose that Ω_c is a compact set. If there is a functional V: $\mathcal{PC}([-\tau, 0], \mathbb{R}^n) \to R$ such that*

1. *There are $\mathcal{K}_\infty$-functions*[1] *α, β such that:*

$$\alpha(\|x(t)\|) \leq V(x_t) \leq \beta(\|x_t\|_{\mathcal{PC}}). \tag{4.61}$$

2. *There is a $\mathcal{K}$-function w such that, for any $\Delta_c \in \Omega_c$ and any nonreset instant t,*

$$W(x_t, \Delta_c) := \dot{V}(x_t) \leq -w(\|x(t)\|). \tag{4.62}$$

3. *The reset mapping $\phi \mapsto \phi^+$ is dissipative with respect to $V(\phi)$ in the sense that for any $\phi \in \mathcal{PC}([-\tau, 0], \mathbb{R}^n)$ satisfying*

$$\phi(0) \in \bigcup_{\Delta_c \in \Omega_c} \ker C_{\Delta_c}, \tag{4.63}$$

there holds

$$V(\phi^+) - V(\phi) \leq 0. \tag{4.64}$$

Here, ker $(\cdot)$ *represents the null space of a matrix.*

Then the RCS (4.56) is asymptotically stable for any admissible uncertainty $\Delta_c(t) \in \Omega_c$.

[1] A continuous function $\alpha : [0, \infty) \to [0, \infty)$ is said to belong to class $\mathcal{K}_\infty$ if it is strictly increasing and satisfies $\alpha(0) = 0$ and $\lim_{r \to \infty} \alpha(r) = \infty$.

Proof: Conditions (1) and (2) imply that $V(x_t)$ is positive definite, radially unbounded, and strictly decreasing at any nonreset time instant. Condition (3) assures that $V(x_t)$ is nonincreasing upon reset actions. Actually, suppose that $t \in [0, \infty)$ is any reset time instant, i.e., $C_{\Delta_c(t)}x(t) = 0$, there must be a $\Delta_c \in \Omega_c$ such that

$$x_t(0) \in \ker C_{\Delta_c} \subset \bigcup_{\Delta_c \in \Omega_c} \ker C_{\Delta_c},$$

which in turn implies that, by (4.63) and (4.64),

$$V(x_t^+) - V(x_t) \leq 0.$$

Since $V(x_t)$ is positive and nonincreasing, there exists a nonnegative number $V_\infty \geq 0$ such that

$$\lim_{t\to\infty} V(x_t) = V_\infty. \tag{4.65}$$

In the following, we prove that $V_\infty = 0$ by contradiction. Assume that $V_\infty > 0$. By (4.65) and monotonicity of $V(x_t)$, there holds

$$0 < V_\infty \leq V(x_t) \leq V(\phi_0), \quad \forall t \geq 0. \tag{4.66}$$

Here, $\phi_0 \in \mathcal{PC}([-\tau, 0], \mathbb{R}^n)$ represents the initial state. Thus, by Condition (1), we have

$$\|x_t\|_{\mathcal{PC}} \geq \beta^{-1}(V_\infty), \quad \forall t \geq 0 \tag{4.67}$$

$$\|x(t)\| \leq \alpha^{-1}(V(\phi_0)), \quad \forall t \geq 0. \tag{4.68}$$

By (4.68), for any $t \geq \tau$, there holds

$$\begin{aligned}\|x_t\|_{\mathcal{PC}} &= \max_{s\in[-\tau,0]} \|x_t(s)\| \\ &= \max_{s\in[-\tau,0]} \|x(t+s)\| \\ &\leq \alpha^{-1}(V(\phi_0)).\end{aligned}$$

Thus there holds, for any $t \geq 0$,

$$0 < \beta^{-1}(V_\infty) \leq \|x_t\|_{\mathcal{PC}} \leq \max\{\alpha^{-1}(V(\phi_0)), \|\phi_0\|_{\mathcal{PC}}\}. \tag{4.69}$$

Inequality (4.67) also implies that there is a sequence $\{t_k\}$ such that

$$t_k \to \infty, \quad k \to \infty,$$

$$\|x(t_k)\| > \beta^{-1}(V_\infty).$$

Note that time-regularization is used to limit the reset rate. For any h satisfying $0 < h < \delta_m$ (δ_m is defined in (4.52)) and for any k, there is a closed time interval I_k such that:

1. $t_k \in I_k$;
2. The length of the interval I_k equals to h;
3. $x(t)$ is continuous within each I_k.

By (4.56), (4.69), and the fact that Ω_c is bounded, there exists a positive number L such that

$$\max_{\Delta_c \in \Omega_c} \{\|A_{\Delta_c} x(t) + E_{\Delta_c} x(t-\tau)\|\} < L, \quad \forall t \in I_k, \ \forall k. \tag{4.70}$$

Equation (4.70) implies that the change rate of $x(t)$ is bounded. By choosing L large enough such that $\beta^{-1}(V_\infty)/(2L) < \delta_m$ and define $h = \beta^{-1}(V_\infty)/(2L)$, we get that, for any $t \in I_k$,

$$\begin{aligned}
\|x(t)\| &= \left\| x(t_k) + \int_{t_k}^{t} [A_{\Delta_c} x(s) + E_{\Delta_c} x(s-\tau)] ds \right\| \\
&\geq \|x(t_k)\| - L|t - t_k| \\
&\geq \beta^{-1}(V_\infty) - Lh \\
&= \frac{1}{2}\beta^{-1}(V_\infty).
\end{aligned}$$

Thus

$$\dot{V}(x_t) \leq -w\left(\frac{1}{2}\beta^{-1}(V_\infty)\right) < 0, \quad t \in I_k.$$

Note that we can always choose a subsequence $\{t_{k_i}\}$ of $\{t_k\}$ such that the intervals I_{k_i} do not overlap. Without losing any generality, we assume that the sequence $\{t_k\}$ is nonoverlapping. For any positive integer M, suppose that there are N_M reset actions within the interval $[0, t_M]$ and $0 \leq s_1 < s_2 < \cdots < s_{N_M} \leq t_M$ are the reset time instants, where N_M is an integer depending on M.

Thus

$$\begin{aligned}
V(x_{t_M}) &= V(\phi_0) + \sum_{i=1}^{N_M} [V(x_{s_i}^{+}) - V(x_{s_i})] + \int_{[0,t_M]} \dot{V}(x_t) dt \\
&\leq V(\phi_0) + \sum_{k=1}^{M-1} \int_{I_k} \dot{V}(x_t) dt \\
&\leq V(\phi_0) - (M-1)hw\left(\frac{1}{2}\beta^{-1}(V_\infty)\right).
\end{aligned} \tag{4.71}$$

Note that $M > 0$ can be chosen arbitrarily. By choosing M such that

$$M > 1 + V(\phi_0)\Big/\left(hw\left(\frac{1}{2}\beta^{-1}(V_\infty)\right)\right)$$

leads to $V(x_{t_M}) < 0$ which contradicts the positiveness of $V(\phi)$.

In the following, we adopt the *Lyapunov functional* candidate of the form

$$V(\phi) = \phi^T(0)P\phi(0) + \int_{-\tau}^{0} \left[\phi^T(\theta)Q\phi(\theta) + \int_{\theta}^{0} \dot{\phi}^T(s)Z\dot{\phi}(s)ds\right] d\theta \tag{4.72}$$

to derive robust stability results, where P, Q, and Z are positive-definite matrices. This kind of functionals is widely adopted to prove stability of reset-free systems with delays [10]. For any solution $x(t)$ to system (4.56), $V(x_t)$ obviously satisfies condition (1) of Proposition 4.7.

Lemma 4.2. *Suppose that the uncertainty set Ω_c contains at least one $\Delta_c = [\delta_1 \ \ \gamma]$ with $\gamma \neq 0$. Then the reset mapping $\phi \mapsto \phi^+$ is dissipative with respect to the functional (4.72) if and only if P is of the form*

$$P = \left[\begin{array}{c|c} P_{11} & \Psi\left[(R_\rho^T - I_\rho)^\perp\right]^T \\ \hline \star & P_{22} \end{array}\right] \tag{4.73}$$

satisfying

$$R_\rho^T P_{22} R_\rho - P_{22} \leq 0, \tag{4.74}$$

where P_{11}, P_{22}, and Ψ are $(n_p + n_r - \rho) \times (n_p + n_r - \rho)$, $\rho \times \rho$, and $(n_p + n_r - \rho) \times (\rho - r_\rho)$ matrices, respectively. $(\cdot)^\perp$ represents right orthogonal complement of matrices and r_ρ represents the rank of $R_\rho - I_\rho$, i.e.,

$$r_\rho = \text{Rank}\,(R_\rho - I_\rho).$$

Proof: Note that

$$V(\phi^+) - V(\phi) = \phi^T(0)(R^T PR - P)\phi(0),$$

the reset mapping $\phi \mapsto \phi^+$ is dissipative with respect to the functional (4.72) if and only if

$$z(R^T PR - P)z \leq 0, \quad \forall z \in \bigcup_{\Delta_c \in \Omega_c} \ker C_{\Delta_c}. \tag{4.75}$$

For any $\Delta_c = [\delta_1, \gamma]$, it is easy to check that one of the right orthogonal complements of C_{Δ_c} is

$$C_{\Delta_c}^\perp = \left[\begin{array}{c|c} \gamma & 0 \\ -(1+\delta_1)I_{n_p-1} & \\ \hline 0 & I_{n_r} \end{array}\right]. \tag{4.76}$$

See Reference 11. Thus by Finsler's Lemma ([12, 13]), (4.75) is equivalent to

$$(C_{\Delta_c}^\perp)^T(R^T PR - P)C_{\Delta_c}^\perp \leq 0, \quad \forall \Delta_c \in \Omega_c. \tag{4.77}$$

Partition $C_{\Delta_c}^\perp$, R, and P as

$$C_{\Delta_c}^\perp = \begin{bmatrix} J_{\Delta_c} & 0 \\ 0 & I_\rho \end{bmatrix}, \quad R = \begin{bmatrix} I_{n_p+n_r-\rho} & 0 \\ 0 & R_\rho \end{bmatrix}, \tag{4.78}$$

$$P = \left[\begin{array}{c|c} P_{11} & P_{12} \\ \hline \star & P_{22} \end{array}\right] \tag{4.79}$$

where I_ρ and $I_{n_p+n_r-\rho}$ are identity matrices with dimensions $\rho \times \rho$ and $(n_p + n_r - \rho) \times (n_p + n_r - \rho)$, respectively, and J_{Δ_c} is an $(n_p + n_r - \rho) \times (n_p + n_r - \rho - 1)$ matrix defined as

$$J_{\Delta_c} = \left[\begin{array}{c|c} \gamma & 0 \\ -(1+\delta_1)I_{n_p-1} & \\ \hline 0 & I_{n_r-\rho} \end{array}\right]. \tag{4.80}$$

By substituting (4.78) and (4.79) into (4.77), it holds that (4.77) is equivalent to

$$\left[\begin{array}{c|c} 0 & J_{\Delta_c}^T P_{12}(R_\rho - I_\rho) \\ \hline \star & R_\rho^T P_{22} R_\rho - P_{22} \end{array}\right] \leq 0, \quad \forall \Delta_c \in \Omega_c$$

which is in turn equivalent to (4.74) and

$$J_{\Delta_c}^T P_{12}(R_\rho - I_\rho) = 0, \quad \forall \Delta_c \in \Omega_c. \tag{4.81}$$

Thus in the following, we only need to prove that (4.81) is equivalent to

$$P_{12} = \Psi \left[(R_\rho^T - I_\rho)^\perp \right]^T$$

for some matrix Ψ. To this end, we partition $P_{12}(R_\rho - I_\rho)$ conformably to that of J_{Δ_c} as

$$P_{12} = \begin{bmatrix} \Gamma_1 \\ \Gamma_2 \\ \Gamma_3 \end{bmatrix}$$

where Γ_1, Γ_2, and Γ_3 are, respectively, $1 \times \rho$, $(n_p - 1) \times \rho$, and $(n_r - \rho) \times \rho$ matrices. Then (4.81) is equivalent to $\Gamma_3 = 0$ and

$$\gamma^T \Gamma_1 - (1 + \delta_1)\Gamma_2 = 0, \quad \forall \Delta_c = [\delta_1 \;\; \gamma] \in \Omega_c. \tag{4.82}$$

By the assumption that Ω_c contains the origin, we can choose $\Delta_c = [\delta_1, \gamma] = 0$ in (4.82) which leads to $\Gamma_2 = 0$ and

$$\gamma^T \Gamma_1 = 0, \quad \forall \Delta_c \in \Omega_c. \tag{4.83}$$

Since Ω_c contains at least one element $\Delta_c = [\delta, \gamma]$ with $\gamma \neq 0$, then (4.83) holds if and only if $\Gamma_1 = 0$. Based on the observation above, (4.81) is equivalent to $P_{12}(R_\rho - I) = 0$, or equivalently, $(R_\rho^T - I)P_{12}^T = 0$. This is in turn equivalent to the existence of a matrix $\tilde{\Psi}$ such that $P_{12}^T = (R_\rho^T - I)^\perp \tilde{\Psi}$. Thus the proof is completed by taking $\Psi = \tilde{\Psi}^T$.

Lemma 4.3. *Suppose that all the elements of Ω_c are of the form $\Delta_c = [\delta_1, \gamma]$ with $\gamma = 0$. Then the reset mapping $\phi \mapsto \phi^+$ is dissipative with respect to the functional (4.72) if and only if the P-matrix in (4.72) is of the form*

$$P = \left[\begin{array}{c|c} P_{11} & \begin{bmatrix} \mu \\ \Pi \left[(R_\rho^T - I_\rho)^\perp \right]^T \end{bmatrix} \\ \hline \star & P_{22} \end{array}\right] \tag{4.84}$$

satisfying (4.74), where μ is a $1 \times \rho$ vector and P_{11}, P_{22}, and Π are $(n_p + n_r - \rho) \times (n_p + n_r - \rho)$, $\rho \times \rho$, and $(n_p + n_r - \rho - 1) \times (\rho - r_\rho)$ matrices, respectively.

Proof: In this case, the output of the system becomes

$$y = (C_p + \Delta_c)x = (1 + \delta_1)x_1.$$

One easily sees that $(1+\delta_1)x_1=0$ if and only if $x_1=0$ provided that $|\delta_1|<1$. This means that the uncertainty δ_1 has no direct impact on the reset time instants. Accordingly, inequality (4.77) becomes

$$(C^{\perp})^T(R^TPR-P)C^{\perp}\leq 0 \tag{4.85}$$

whose left-hand side is independent of the uncertainty Δ_c. By taking

$$C^{\perp}=\begin{bmatrix}0\\ I_{n_p+n_r-1}\end{bmatrix}$$

and following the same argument of Lemma 4.2, the above inequality is equivalent to (4.74) and

$$J^TP_{12}(R_\rho-I_\rho)=0 \tag{4.86}$$

with

$$J=\begin{bmatrix}0_{1\times(n_p+n_r-\rho-1)}\\ I_{n_p+n_r-\rho-1}\end{bmatrix}.$$

Partition P_{12} as

$$P_{12}=\begin{bmatrix}\mu\\ \Gamma\end{bmatrix}$$

where μ is a $1\times\rho$ vector and Γ is a $(n_p+n_r-\rho-1)\times\rho$ matrix. Thus (4.86) is equivalent to

$$\Gamma(R_\rho-I_\rho)=0$$

which implies that there is a $(n_p+n_r-\rho-1)\times(\rho-r_\rho)$ matrix Π such that

$$\Gamma=\Pi\left[(R_\rho^T-I_\rho)^{\perp}\right]^T.$$

Remark 4.9. *The reset matrix R_r is in general assumed to be of the form (4.54) with $R_\rho=0$ in the literature. In Reference 11, R_ρ is assumed to be Schur stable, i.e., $\lambda(R_\rho)<1$. In this case,* Rank $(R_\rho^T-I_\rho)=\rho$ *and $(R_\rho^T-I_\rho)^{\perp}=0$. Thus the P-matrices in (4.73) and (4.84) degenerate to*

$$P=\left[\begin{array}{c|c}P_{11}&0\\ \hline 0&P_{22}\end{array}\right]$$

and

$$P=\left[\begin{array}{c|c}P_{11}&\begin{bmatrix}\mu\\ 0\end{bmatrix}\\ \hline \star&P_{22}\end{array}\right],$$

respectively. Here, we do not assume Schur stability of R_ρ, thus the case that $R_\rho-I_\rho$ being singular is also included. Especially, when $R_\rho-I_\rho=0$, we have Rank $((R_\rho^T-I_\rho)^{\perp})=\rho$ *which implies that the identity matrix I_ρ is one of the choices* *for $(R_\rho^T - I_\rho)^\perp$. In this case, (4.74) naturally holds and the P-matrices in both (4.73) and (4.84) become positive-definite matrices without any structural constraints, i.e., the case without reset is naturally included in Lemma 4.2.*

Proposition 4.8. *Suppose that the uncertainty set is of the form*

$$\Omega_c = \overline{co}\{\Delta_1, \ldots, \Delta_N\}$$

which contains at least one $\Delta_c = [\delta_1 \ \ \gamma]$ with $\gamma \neq 0$. If there exist positive definite matrices P_{11}, P_{22}, Q, and Z whose dimensions are $(n_p + n_r - \rho) \times (n_p + n_r - \rho)$, $\rho \times \rho$, $(n_p + n_r) \times (n_p + n_r)$, and $(n_p + n_r) \times (n_p + n_r)$, respectively, matrices X_{11}, X_{12} X_{22}, Y, and U whose dimensions are all $(n_p + n_r) \times (n_p + n_r)$, a $(n_p + n_r - \rho) \times \rho$ matrix Ψ such that

$$\Phi(\Delta_i) := \left[\begin{array}{cc|c} \Phi_{11}(\Delta_i) & \Phi_{12}(\Delta_i) & \tau A_{\Delta_i}^T Z \\ \star & \Phi_{22} & \tau E_{\Delta_i}^T Z \\ \hline \star & \star & -\tau Z \end{array}\right] < 0, \tag{4.87}$$

$\forall i \in \{1, \ldots, N\}$,

$$\Psi := \begin{bmatrix} X_{11} & X_{12} & Y \\ \star & X_{22} & U \\ \star & \star & Z \end{bmatrix} \geq 0 \tag{4.88}$$

$$R_\rho^T P_{22} R_\rho - P_{22} \leq 0 \tag{4.89}$$

where

$$\begin{aligned} \Phi_{11}(\Delta_i) &= Q + \tau X_{11} + Y + Y^T + \mathrm{Sym}\left\{\left[\begin{array}{c|c} P_{11} & \Psi\left[(R_\rho^T - I_\rho)^\perp\right]^T \\ \star & P_{22} \end{array}\right] A_{\Delta_i}\right\}, \\ \Phi_{12}(\Delta_i) &= -Y + U^T + \tau X_{12} + \left[\begin{array}{c|c} P_{11} & \Psi\left[(R_\rho^T - I_\rho)^\perp\right]^T \\ \star & P_{22} \end{array}\right] E_{\Delta_i}, \\ \Phi_{22} &= -\mathrm{Sym}\{U\} - Q + \tau X_{22}, \end{aligned}$$

where Sym$(\cdot)$ *represents the symmetrization of a matrix, i.e.,* Sym$(\cdot) = (\cdot) + (\cdot)^T$. *Then the RCS (4.56) is asymptotically stable for any $\Delta_c(t) \in \Omega_c$.*

Proof: Take (4.72) as the Lyapunov functional candidate. By following the same argument in Reference 10,

$$\dot{V}(x_t) \leq \xi(t)^T \Xi(\Delta_c)\xi(t) - \int_{-\tau}^{0} \zeta^T(t,s)\Psi\zeta(t,s)ds \tag{4.90}$$

where

$$\xi = \begin{bmatrix} x_t(0) \\ x_t(-\tau) \end{bmatrix}, \quad \zeta(t,s) = \begin{bmatrix} x_t(0) \\ x_t(-\tau) \\ \dot{x}_t(s) \end{bmatrix}$$

and

$$\Xi(\Delta_c) = \begin{bmatrix} \Phi_{11}(\Delta_c) + \tau A_{\Delta_c}^T Z A_{\Delta_c} & \Phi_{12}(\Delta_c) + \tau A_{\Delta_c}^T Z E_{\Delta_c} \\ \star & \Phi_{22}(\Delta_c) + \tau E_{\Delta_c}^T Z E_{\Delta_c} \end{bmatrix}.$$

By the affine dependence of $\Phi(\Delta_c)$ on Δ_c and the convexity of Ω_c, (4.87) implies that

$$\Phi(\Delta_c) < 0, \quad \forall \Delta_c \in \Omega_c. \tag{4.91}$$

Applying the Schur complement, (4.91) implies that

$$\Xi(\Delta_c) < 0, \quad \forall \Delta_c \in \Omega_c.$$

By compactness of Ω_c and continuity of $\Xi(\Delta_c)$ with respect to Δ_c, there is a positive number λ such that

$$\Xi(\Delta_c) < -\lambda I, \quad \forall \Delta_c \in \Omega_c.$$

Thus

$$\dot{V}(x_t) \le -\lambda \|x_t(0)\|^2$$

and condition (2) of Proposition 4.7 is satisfied. At last, by Lemma 4.2, condition (3) of Proposition 4.7 holds. Thus the RCS (4.56) is asymptotically stable for any admissible $\Delta_c(t) \in \Omega_c$.

The following result can be similarly proved by using Lemma 4.3.

Proposition 4.9. *Suppose that the uncertainty set is of the form $\Omega_c = \overline{co}\{\Delta_1, \ldots, \Delta_N\}$ and all the elements of Ω_c are of the form $\Delta_c = [\delta_1, \gamma]$ with $\gamma = 0$. If there exist positive definite matrices P_{11}, P_{22}, Q, and Z whose dimensions are $(n_p + n_r - \rho) \times (n_p + n_r - \rho)$, $\rho \times \rho$, $(n_p + n_r) \times (n_p + n_r)$, and $(n_p + n_r) \times (n_p + n_r)$, respectively, matrices X_{11}, X_{12} X_{22}, Y, and U whose dimensions are all $(n_p + n_r) \times (n_p + n_r)$, a $1 \times \rho$ vector μ, and a $(n_p + n_r - \rho - 1) \times (\rho - r_\rho)$ matrix Π, such that (4.87), (4.88), and (4.89) hold with*

$$\Phi_{11}(\Delta_i) = Q + \tau X_{11} + Y + Y^T + \mathrm{Sym}\left\{ \left[\begin{array}{c|c} P_{11} & \begin{bmatrix} \mu \\ \Pi \left[(R_\rho^T - I_\rho)^\perp \right]^T \end{bmatrix} \\ \hline \star & P_{22} \end{array} \right] A_{\Delta_i} \right\},$$

$$\Phi_{12}(\Delta_i) = -Y + U^T + \tau X_{12} + \left[\begin{array}{c|c} P_{11} & \begin{bmatrix} \mu \\ \Pi \left[(R_\rho^T - I_\rho)^\perp \right]^T \end{bmatrix} \\ \hline \star & P_{22} \end{array} \right] E_{\Delta_i},$$

$$\Phi_{22} = -\mathrm{Sym}\{U\} - Q + \tau X_{22},$$

then the RCS (4.56) is asymptotically stable for any $\Delta_c(t) \in \Omega_c$.

Remark 4.10. *Note that the left-hand side of (4.87) is linearly dependent on the delay τ, thus if the conditions of Proposition 4.8 are satisfied for both τ_0 and τ_1, $\tau_0 < \tau_1$, then they are satisfied for any constant $\tau \in [\tau_0, \tau_1]$ as well.*

Remark 4.11. *Different or even less conservative results could be obtained by combining Lemmas 4.2 and 4.3 with the techniques developed in the literature, including Jensen's inequality [7] and time-delay discretization [14] and uncertain Finsler's Lemma ([12, 13]), etc.*

4.5 Examples

Example 4.1. *Consider a system consisting of the plant*

$$\Sigma_p(\gamma): \quad \begin{cases} \dot{x}_p = \begin{bmatrix} -0.6 & -0.4 \\ 1 & 0 \end{bmatrix} x_p + \begin{bmatrix} 2 \\ 0 \end{bmatrix} u \\ y = \begin{bmatrix} 0.5 & 0.5+\gamma \end{bmatrix} x_p \end{cases} \tag{4.92}$$

and the reset controller

$$\Sigma_R: \quad \begin{cases} \dot{x}_r = -x_r - y, & y \neq 0 \\ x_r^+ = r x_r, & y = 0, \\ u = x_r, \end{cases} \tag{4.93}$$

where r is a scalar satisfying $|r| < 1$. In the above system, γ is an uncertain parameter. Denote $\Delta_c = [0, \gamma]$ and define a transformation as

$$z_p = \begin{bmatrix} 0.5 & 0.5 \\ 0 & 1 \end{bmatrix} x_p,$$

then the plant becomes

$$\tilde{\Sigma}_p(\gamma): \quad \begin{cases} \dot{z}_p = \begin{bmatrix} 0.4 & -0.4 \\ 2 & -1 \end{bmatrix} z_p + \begin{bmatrix} 1 \\ 0 \end{bmatrix} u \\ y = \begin{bmatrix} 1 & \gamma \end{bmatrix} z_p. \end{cases} \tag{4.94}$$

It is trivial to check that the closed-loop system is of the form (4.5) with $x = [z_p^T, x_r^T]^T$, $C_\Delta = \begin{bmatrix} 1 & \gamma & 0 \end{bmatrix}$ and

$$A_\Delta = \begin{bmatrix} 0.4 & -0.4 & 1 \\ 2 & -1 & 0 \\ -1 & -\gamma & -1 \end{bmatrix} := A(\gamma), \quad R = \begin{bmatrix} 1 & & \\ & 1 & \\ & & r \end{bmatrix}.$$

Since $r \in (-1, 1)$, according to Proposition 4.3, this RCS is quadratically stable with respect to time-varying uncertainty $\gamma(t) \in [\underline{\delta}, \bar{\delta}]$ with $\underline{\delta} < 0$ and $\bar{\delta} > 0$ if and only if there is a block-diagonal positive matrix

$$P = \begin{bmatrix} P_{11} & \\ & p_{22} \end{bmatrix}$$

where P_{11} is a 2×2 positive definite matrix and p_{22} is a positive scalar number such that

$$A(\gamma)^T P + PA(\gamma) < 0, \quad \gamma = \underline{\delta} \text{ and } \bar{\delta}, \tag{4.95}$$

i.e., P is a common Lyapunov function for $A(\underline{\delta})$ and $A(\bar{\delta})$. Note that we do not need to check (4.24) since it is equivalent to $r^2 - 1 \leq 0$ which is naturally satisfied. By Matlab LMI toolbox, it is easy to check that (4.95) holds with $\underline{\delta} = -0.3$, $\bar{\delta} = 0.4$,

$$P_{11} = \begin{bmatrix} 2.1702 & -0.5721 \\ \star & 0.4683 \end{bmatrix}, \quad p_{22} = 1.3525.$$

Thus this RCS is quadratically stable with respect to $\gamma(t) \in [-0.3, 0.4]$. However, when the interval extends to $[-0.4, 0.45]$, the conditions of Proposition 4.3 fail to satisfy. Thus this RCS is not quadratically stable with respect to this interval.

In the following, we use Proposition 4.6 to check the affine quadratic stability for constant uncertainty $\gamma \in \Omega_\delta = [-0.7, 0.59]$. It is trivial to check that $A(\gamma)$ with $\gamma = \frac{0.59-(-0.7)}{2}$ is stable. Note that the uncertainty γ only enters the second element of C, there holds $\delta_1 = 0$. Thus P_1 can be chosen as a zero matrix. By Matlab LMI toolbox, it is easy to check that the conditions of Proposition 4.6 hold with $\mu = -2.1127$ and

$$\Gamma_{11}^0 = \begin{bmatrix} 2.8736 \times 10^3 & -8.549 \times 10^2 \\ \star & 7.456 \times 10^2 \end{bmatrix}, \quad \Gamma_{22}^0 = 1.6579 \times 10^3,$$

$$\Gamma_{11}^1 = \begin{bmatrix} 1.2814 \times 10^3 & -2.843 \times 10^2 \\ \star & 4.547 \times 10^2 \end{bmatrix}, \quad \Gamma_{22}^1 = 0.$$

Thus this RCS is affinely quadratically stable for any $\gamma \in [-0.7, 0.59]$ with a parameter-dependent Lyapunov function

$$P(\gamma) = P_0 + \gamma P_2.$$

If we extend the interval to $[-0.7, 0.6]$, the conditions of Proposition 4.6 fail to hold.

Some simulation results are given in Figures 4.1–4.3 to show robustness of the RCS. In the simulation, $r = -0.5$ and the reset rate is limited by time-regularization with $\tau_{\min} = 0.2$. See Remark 4.1. Figure 4.1 gives the state responses $x_p = (x_{p1}, x_{p2})^T$ and output responses for uncertainty-free case. It shows that reset action improves the convergence rate. Figure 4.2 gives the responses for time-varying uncertainty $\gamma(t) = 0.2 \sin(20t)$. One sees that even though the uncertainties deteriorate the performance, the controller with reset still outperforms the one without reset in this case. In Figure 4.3, the responses for different γ are given.

For the constant uncertainty case, by searching over the interval $\gamma \in [-1, 1]$, the exact region for quadratic stability is $[-0.7, 0.9]$, i.e., for any constant $\gamma \in [-0.7, 0.9]$, there is a quadratic Lyapunov function for the baseline system which is also dissipative with respect to reset actions. This indicates the conservativeness of Proposition 4.6. In Figure 4.3, the responses for $\gamma = 0.6$ and 0.9 are also included.

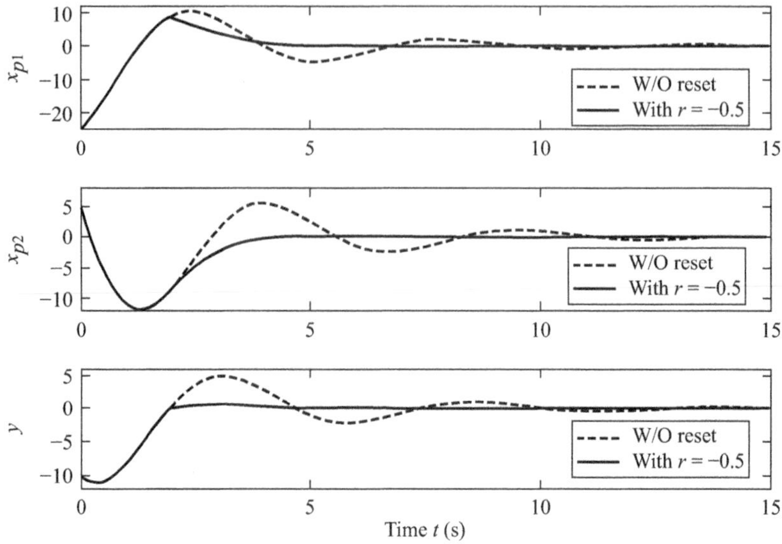

Figure 4.1 Responses for the uncertainty-free case

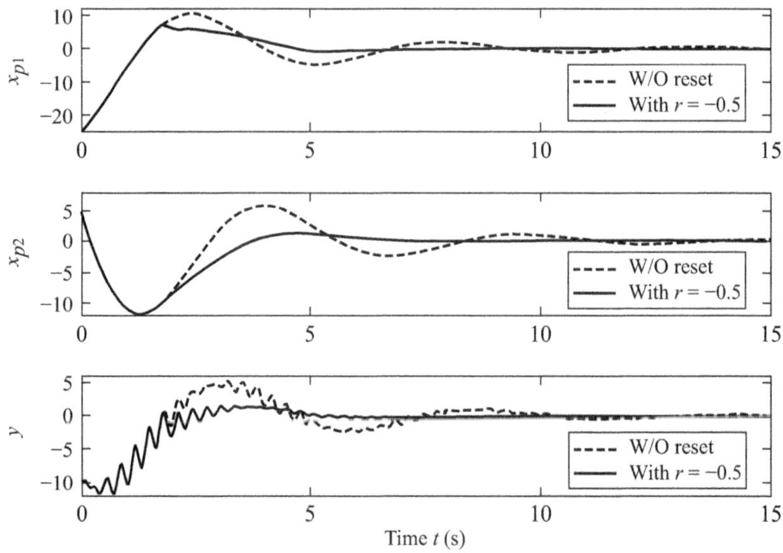

Figure 4.2 Responses for uncertainty $\gamma(t) = 0.2\sin(20t)$

Example 4.2. *Consider a system consisting of the plant*

$$\Sigma_p(\gamma): \quad \begin{cases} \dot{x}_p = \begin{bmatrix} -0.6 & -0.4 \\ 1 & 0 \end{bmatrix} x_p + \begin{bmatrix} 2 \\ 0 \end{bmatrix} u_p \\ y_p = \begin{bmatrix} 0.5 & 0.5 + \gamma \end{bmatrix} x_p \end{cases} \tag{4.96}$$

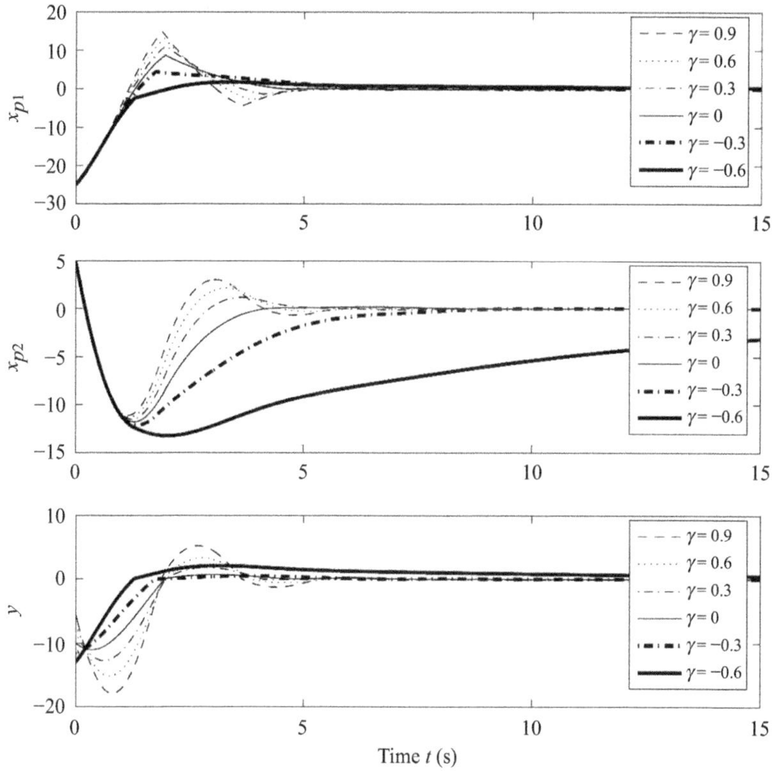

Figure 4.3 Responses for constant uncertainty γ ($r = -0.5$)

and the reset controller

$$\Sigma_R : \begin{cases} \dot{x}_r = -x_r + u_r, & u_r \neq 0 \text{ or } \delta < \delta_m \\ x_r^+ = r x_r, & u_r = 0 \text{ and } \delta \geq \delta_m, \\ y_r = x_r, \end{cases} \tag{4.97}$$

which are interconnected via (4.55) where r is a scalar satisfying $|r| < 1$. In the above system, $\gamma \in [\gamma_l, \gamma_r]$ is an uncertain parameter, $\gamma_l < 0 < \gamma_r$. Denote $\Delta_c = [0, \gamma]$ and define a transformation as

$$z_p = \begin{bmatrix} 0.5 & 0.5 \\ 0 & 1 \end{bmatrix} x_p,$$

then the plant becomes

$$\tilde{\Sigma}_p(\gamma) : \begin{cases} \dot{z}_p = \begin{bmatrix} 0.4 & -0.4 \\ 2 & -1 \end{bmatrix} z_p + \begin{bmatrix} 1 \\ 0 \end{bmatrix} u_p \\ y = \begin{bmatrix} 1 & \gamma \end{bmatrix} z_p. \end{cases} \tag{4.98}$$

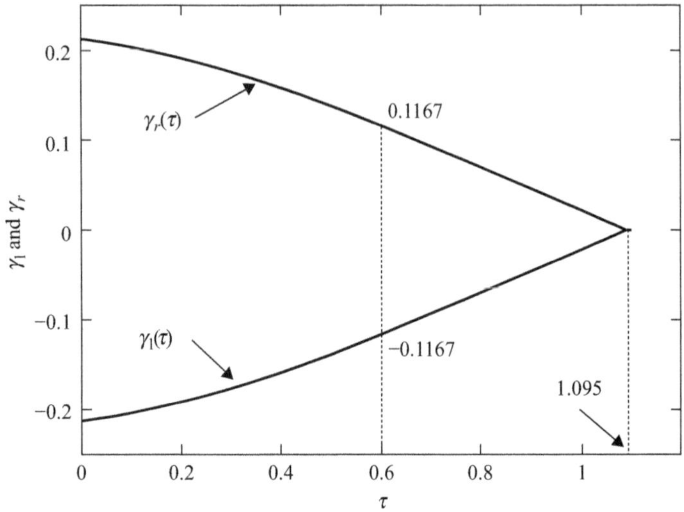

Figure 4.4 *The curves of γ_l and γ_r over τ. (For any given $\tau \in [0, 1.095]$, the RCS is robustly stable respect to time-varying uncertainty $\gamma(t) \in [\gamma_l(\tau), \gamma_r(\tau)]$.)*

It is trivial to check that the closed-loop system is of the form (4.56) with $x = [z_p^T, x_r^T]^T$, $C_{\Delta_c} = \begin{bmatrix} 1 & \gamma & 0 \end{bmatrix}$ *and*

$$A_{\Delta_c} = \begin{bmatrix} 0.4 & -0.4 & 1 \\ 2 & -1 & 0 \\ -1 & -\gamma & -1 \end{bmatrix} := A(\gamma),$$

$$E_{\Delta_c} = \begin{bmatrix} 0 & 0 & 1 \\ 0 & 0 & 0 \\ 0 & 0 & 0 \end{bmatrix}, \quad R = \begin{bmatrix} 1 & & \\ & 1 & \\ & & r \end{bmatrix}.$$

Since $|r| < 1$, we have $\rho = 1$ and $R_\rho - I_\rho = 1$. Thus we can choose $[(R_\rho^T - I_\rho)^\perp]^T = 0$, i.e., the off-diagonal block P_{12} of matrix P is zero. Proposition 4.8 is used to test the robust stability. By using Matlab LMI tool, one easily sees that the conditions of Proposition 4.8 are feasible only if $\tau \in [0, 1.095]$. For any given $\tau \in [0, 1.095]$, the largest symmetric interval $[\gamma_l(\tau), \gamma_r(\tau)]$ for which the conditions of Proposition 4.8 hold is shown in Figure 4.4. For instance, from this figure, when $\tau = 0.6$, the RCS is robustly stable with respect to uncertainty $\gamma(t) \in [-0.1167, 0.1167]$.

Figure 4.5 gives the initial condition response with $x(0) = [-2 \;\; 2]^T$ for $\tau = 0.6$, $\gamma = 0.1 \sin 15t$ and $\delta_m = 1$. Note that in this case, the system is asymptotically stable for any $r \in (-1, 1)$, thus we can safely adjust r to improve the time response. The dash curve is for the case without reset. The dash dot and the solid curves are for the cases with $r = 0$ and $r = -0.7$, respectively. Figure 4.6 gives the control input response u_p.

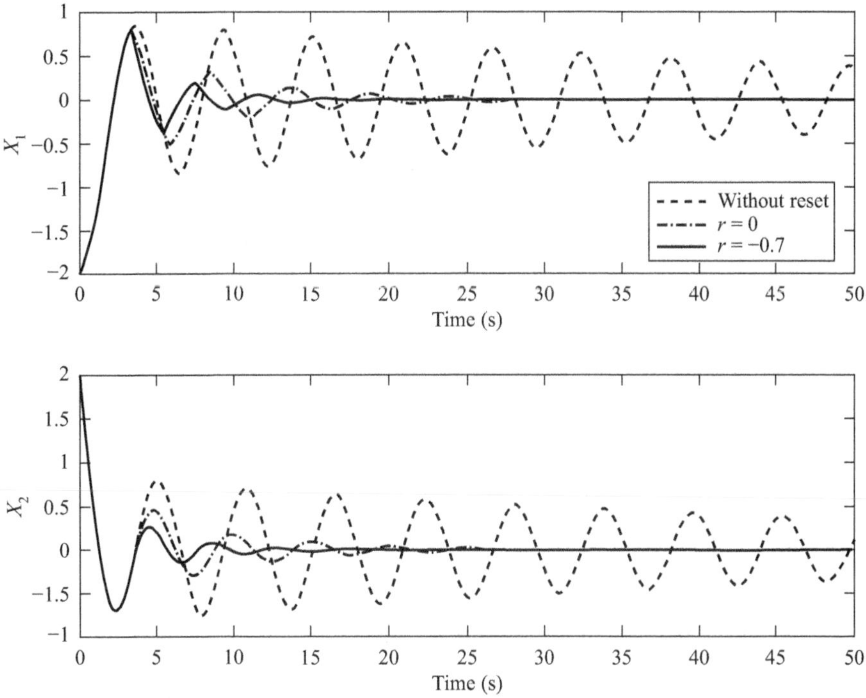

Figure 4.5 Initial condition response to $x(0) = [-2\ \ 2]^T$

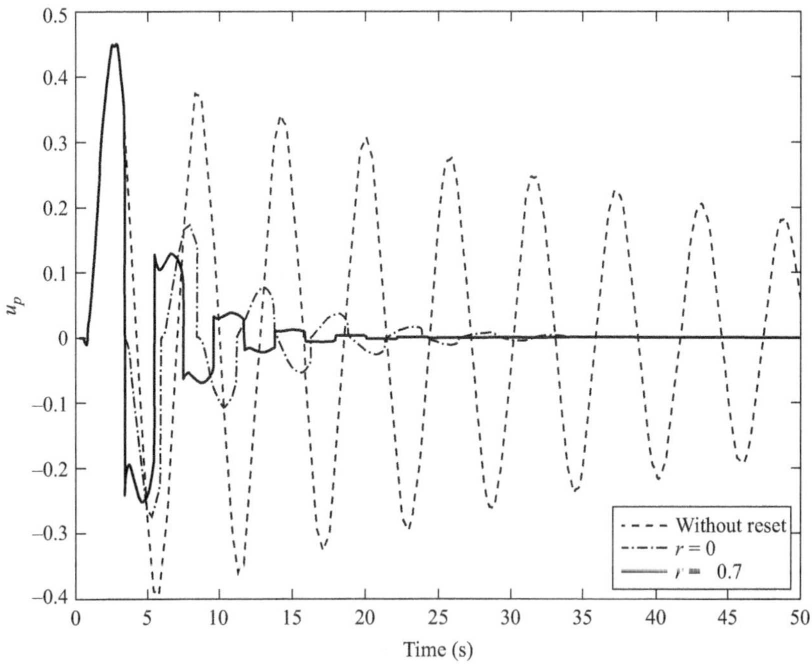

Figure 4.6 Input u_p

Notes

The materials regarding quadratic stability and affine quadratic stability are mainly based on References 11 and 15. The results regarding quadratic stability of RCSs with delays can be found in Reference 16. The technique of full-rank annihilators of matrices has been applied in Reference 8 to treat quadratic stability of RCSs without uncertainties. For quadratic stability of reset systems without uncertainty and affine quadratic stability of linear systems without reset, we refer the reader to References 8 and 5.

References

[1] D. Nešic, L. Zaccarian, and A.R. Teel. Stability properties of reset systems. *Automatica*, 44(8):2019–2026, 2008.

[2] J. Carrasco, A. Baños, and A. van der Schaft. A passivity-based approach to reset control systems stability. *Systems & Control Letters*, 59(1):18–24, 2010.

[3] F. Amato. Robust control of linear systems subject to uncertain time-varying parameters. In *Lecture Notes in Control and Information Sciences*, Springer Berlin Heidelberg, 2006.

[4] S. Boyd, L. El Ghaoul, E. Feron, and V. Balakrishnan. Linear matrix inequalities in system and control theory. In *SIAM Studies in Applied Mathematics*, vol. 15, Society for Industrial and Applied Mathematics, Philadelphia, Pennsylvania, 1994.

[5] P. Gahinet, P. Apkarian, and M. Chilali. Affine parameter-dependent Lyapunov functions and real parametric uncertainty. *IEEE Transactions on Automatic Control*, 41(3):436–442, 1996.

[6] A. Baños and A. Barreiro. Delay-independent stability of reset systems. *IEEE Transactions on Automatic Control*, 54(2):341–346, 2009.

[7] K. Gu, V. Kharitonov, and J. Chen. Stability of time-delay systems. In *Control Engineering*. Birkhäuser, Boston, Basel, Berlin, 2003.

[8] O. Beker, C.V. Hollot, Y. Chait, and H. Han. Fundamental properties of reset control systems. *Automatica*, 40(6):905–915, 2004.

[9] A. Barreiro and A. Baños. Delay-dependent stability of reset systems. *Automatica*, 46(1):216–221, 2010.

[10] M. Wu, Y. He, J.H. She, and G.P. Liu. Delay-dependent criteria for robust stability of time-varying delay systems. *Automatica*, 40(8):1435–1439, 2004.

[11] Y. Guo, Y. Wang, and L. Xie. Robust stability of reset control systems with uncertain output matrix. *Automatica*, 48(8):1879–1884, 2012.

[12] R.E. Skelton, T. Iwasaki, and D.E. Grigoriadis. A unified algebraic approach to control design. In *Series in Systems and Control*. Taylor & Francis Ltd. London, UK, 1997.

[13] T. Iwasaki. LPV system analysis with quadratic separator. *The 37th IEEE Conference on Decision and Control*, vol. 3. IEEE, Tampa, Florida, pp. 3021–3026, 1998.

[14] F. Gouaisbaut, D. Peaucelle, et al. Delay-dependent robust stability of time delay systems. *5th IFAC Symposium on Robust Control Design*, July 2006, Toulouse, France. IFAC, 12p.

[15] Y. Guo, W. Gui, and C. Yang. Quadratic stability of uncertain reset control systems. *The 18th IFACWorld Congress*, August 28–September 2, 2011, IFAC, Milano, Italy, pp. 6297–6300, 2011.

[16] Y. Guo and L. Xie. Quadratic stability of reset control systems with delays. *10th World Congress on Intelligent Control and Automation*, July 2012, IEEE, Beijing, pp. 2268–2273, 2012.

Chapter 5

RCSs with discrete-time reset conditions

Note that the input-dependent reset mechanism brings possibility of Zeno behavior and beating phenomenon [1, 2]. One approach to avoid these phenomena is to limit the reset rate by *time-regularization* [1, 3]. Another way is to sample the triggering signal at a pre-specified rate and reset the controller states only when the discretized triggering signal crosses zero. Precisely, let the tracking error $e(t)$ be the triggering signal and T_s the sampling period. Denote $e_k = e(kT_s)$ for simplicity. Then the controller state is reset at $t = kT_s$ whenever

$$e_k e_{k-1} \leq 0. \tag{5.1}$$

In computer-based implementation, due to sampling rate limitation of physical devices, the exact time instants of the zero-crossings of a continuous signal are difficult to be detected. Thus the triggering condition has to be replaced by a discrete-time counterpart. When the sampling rate is much lower than the natural frequency of the baseline system, the sampling would have significant effect on the performance. Based on this observation, it is imperative to analyze and design *reset control systems* (RCSs) based on discrete-time triggering conditions directly.

In this chapter, we consider the RCSs whose triggering conditions are of the form

$$\phi(e_k, e_{k-1}) \leq 0 \tag{5.2}$$

where $\phi(\alpha, \beta)$ is a quadratic function called the *triggering function*. Note that (5.1) is a special case of (5.2). Our first goal is to establish stability criteria for RCSs with the discrete-time triggering condition. The basic idea is to transfer the stability problem of an RCS to that of a switched system and then many tools, which were originally developed for switched systems, can be applied. Our second goal is to reveal that a well-designed RCS can be regarded as a hybrid system combining an underdamped baseline system with an overdamped reset mode. In addition, it will be shown that the reset matrix has significant effect on the damping of the reset mode. This indicates that the reset matrix may be adjusted to further improve the performance. Based on this observation, we then propose a guideline of reset controller design for *single-input single-output* (SISO) systems to achieve good transient performance. It includes three steps: baseline controller design, reset matrix design, and triggering condition tuning. The design of the baseline controller is similar to that of the Horowitz's method for which the main principle is to meet other specifications at the expense of violating the overshoot constraint. Then, the reset matrix is adjusted such that the reset mode is overdamped. The last step is to tune the triggering condition so that the baseline system and the reset mode cooperate well with each other. In order to demonstrate the design process and to show the validity of the stability criteria obtained, a *hard disk drive* (HDD) example is considered. From this example, one sees that the tunable reset matrix and triggering condition provide additional freedoms in the control design.

5.1 Preliminaries and problem setting

Consider a continuous plant

$$\begin{cases} \dot{x}_p = A_p x_p + B_p u \\ y = C_p x_p \end{cases} \tag{5.3}$$

with $x_p \in \mathbb{R}^{n_p}$, $u \in \mathbb{R}^{n_i}$, $y \in \mathbb{R}^{n_o}$, and a reset controller

$$\begin{cases} \dot{z} = A_r z + B_r y + K_r r, & t \neq kT_s \\ z_k^+ = z_k, & \phi(e_k, e_{k-1}) > 0 \\ z_k^+ = R_r z_k + J_r y_k + L_r r, & \phi(e_k, e_{k-1}) \leq 0 \\ u = F_r z + G_r y + H_r r \end{cases} \tag{5.4}$$

with $z \in \mathbb{R}^{n_r}$ the controller state, $r \in \mathbb{R}^{n_o}$ the constant reference input, and $e = r - y$ the tracking error. e_k, z_k, and y_k are short for $e(kT_s)$, $z(kT_s)$, and $y(kT_s)$, respectively, with T_s the sampling period and $e_{-1} := e_0$. $z(t)$ is assumed to be continuous from the left and z_k^+ represents the controller state after reset at $t = kT_s$, i.e.,

$$z_k^+ = \lim_{t \to kT_s + 0} z(t).$$

R_r is the *reset matrix*. The triggering function $\phi(\alpha, \beta)$ is a mapping from $\mathbb{R}^{n_o} \times \mathbb{R}^{n_o}$ to $\mathbb{R}$. In this chapter, we consider a quadratic triggering function of the form

$$\phi(\alpha, \beta) = \begin{bmatrix} \alpha^T & \beta^T \end{bmatrix} S \begin{bmatrix} \alpha \\ \beta \end{bmatrix}, \tag{5.5}$$

where S is a $2n_o \times 2n_o$ symmetric matrix. For instance, for single output systems, the triggering condition (5.1) can be represented as $\phi(e_k, e_{k-1}) \leq 0$ with

$$S = \begin{bmatrix} 0 & 1/2 \\ 1/2 & 0 \end{bmatrix}. \tag{5.6}$$

In some cases, we may want the reset time instants to be earlier than the zero-crossings. In order to achieve this, we can use the historical data to predict whether the tracking error will cross zero in some pre-specified period of time. For instance, a simple linear predictor of the tracking error for $t = kT_s + \Delta$ is given by

$$\hat{e}(kT_s + \Delta) = \frac{e_k - e_{k-1}}{T_s} \Delta + e_k$$

where Δ is a pre-specified positive constant. Then reset occurs at $t = kT_s$ if

$$\left[\frac{e_k - e_{k-1}}{T_s} \Delta + e_k \right] e_{k-1} \leq 0. \tag{5.7}$$

Δ roughly characterizes how much time the reset action is ahead of the zero-crossing and can be a tunable parameter in design. Actually, the triggering condition (5.7) can be also represented as $\phi(e_k, e_{k-1}) \leq 0$ with

$$S = \begin{bmatrix} 0 & (\Delta/T_s + 1)/2 \\ (\Delta/T_s + 1)/2 & -\Delta/T_s \end{bmatrix}. \tag{5.8}$$

Note that the triggering condition considered in this chapter also applies to multi-output systems. For instance, a vector version of (5.1) can be $e_k^T e_{k-1} \leq 0$. In this case, the triggering condition can be represented as $\phi(e_k, e_{k-1}) \leq 0$ with

$$S = \begin{bmatrix} 0 & 1/2 \\ 1/2 & 0 \end{bmatrix} \otimes I.$$

Here, $\otimes$ represents the Kronecker product of matrices.

Combining (5.3) with (5.4) leads to the closed-loop system

$$\begin{cases} \dot{x} = Ax + Kr, & t \neq kT_s \\ x_k^+ = x_k, & \phi(e_k, e_{k-1}) > 0 \\ x_k^+ = Rx_k + Lr, & \phi(e_k, e_{k-1}) \leq 0 \\ y = Cx \end{cases} \tag{5.9}$$

with $x = (x_p^T, z^T)^T$ and

$$A = \begin{bmatrix} A_p + B_p G_r C_p & B_p F_r \\ B_r C_p & A_r \end{bmatrix}, \quad K = \begin{bmatrix} B_p H_r \\ K_r \end{bmatrix}, \tag{5.10}$$

$$R = \begin{bmatrix} I & 0 \\ J_r C_p & R_r \end{bmatrix}, \quad L = \begin{bmatrix} 0 \\ L_r \end{bmatrix}, \quad C = \begin{bmatrix} C_p & 0 \end{bmatrix}. \tag{5.11}$$

The goal of the reset of this chapter is to propose a heuristic guideline of designing a regulator of the form (5.4) which renders the output y to asymptotically track the reference input r, i.e., $\lim_{t\to\infty} y(t) = r$, with good transient performance. The following assumption allows us to transfer the asymptotic tracking property to asymptotic stability.

Assumption 5.1. *For any constant reference r, the baseline system*

$$\begin{cases} \dot{x} = Ax + Kr \\ y = Cx \end{cases}$$

and the reset mapping

$$x^+ = Rx + Lr$$

share a common equilibrium point $\mathbf{x}(r)$ *in the sense that*

$$\begin{cases} A\mathbf{x}(r) + Kr = 0 \\ (R - I)\mathbf{x}(r) + Lr = 0 \\ C\mathbf{x}(r) - r = 0. \end{cases} \tag{5.12}$$

Under Assumption 5.1, define

$$\xi = x - \mathbf{x}(r).$$

Then the closed loop-system (5.9) is transformed to

$$\begin{cases} \dot{\xi} = A\xi, & t \neq kT_s \\ \xi_k^+ = \xi_k, & \phi(C\xi_k, C\xi_{k-1}) > 0 \\ \xi_k^+ = R\xi_k, & \phi(C\xi_k, C\xi_{k-1}) \leq 0 \end{cases} \tag{5.13}$$

where ξ_k is short for $\xi(kT_s)$ and $\xi_{-1} := \xi_0$.

Proposition 5.1. *Suppose that Assumption 5.1 holds. If the reset system (5.13) is globally asymptotically stable, then under the reset controller (5.4), there holds*

$$\lim_{t\to\infty} y(t) = r$$

for any given constant reference input r.

Proof: It follows directly from the fact that

$$e(t) := r - y = C\mathbf{x}(r) - y = -C\xi \to 0 \quad (t \to \infty).$$

5.2 Stability analysis

In this section, we consider the stability of system (5.13). By the special structure of the matrices C and R in (5.11), there holds

$$C(R - I) = 0,$$

which yields

$$C\xi_k = C\xi_k^+, \quad \forall k.$$

This coincides with the fact that only the controller states are reset and thus the output of the system is still continuous. Define $\eta_k = \xi_k^+$ and

$$Q = \begin{bmatrix} e^{A^T T_s} C^T & C^T \end{bmatrix} S \begin{bmatrix} Ce^{AT_s} \\ C \end{bmatrix}. \tag{5.14}$$

Then,

$$\begin{aligned} \phi(C\xi_k, C\xi_{k-1}) &= \phi(C\xi_k, C\xi_{k-1}^+) \\ &= \phi(Ce^{AT_s}\xi_{k-1}^+, C\xi_{k-1}^+) \\ &= \eta_{k-1}^T Q \eta_{k-1}. \end{aligned}$$

When $\eta_k^T Q \eta_k > 0$, no reset occurs at $t = (k+1)T_s$. In this case,

$$\eta_{k+1} = \xi_{k+1}^+ = \xi_{k+1} = e^{AT_s}\xi_k^+ = e^{AT_s}\eta_k.$$

When $\eta_k^T Q \eta_k \leq 0$, reset occurs at $t = (k+1)T_s$ and

$$\eta_{k+1} = \xi_{k+1}^+ = R\xi_{k+1} = Re^{AT_s}\xi_k^+ = Re^{AT_s}\eta_k.$$

Thus the evolution of η_k is governed by the switched system

$$\eta_{k+1} = A_{\sigma(\eta_k)}\eta_k \tag{5.15}$$

where $\sigma(\eta)$ is the switching law defined by

$$\sigma(\eta) = \begin{cases} 1, & \eta^T Q\eta > 0 \\ 2, & \eta^T Q\eta \le 0 \end{cases} \tag{5.16}$$

and A_σ is defined by

$$A_\sigma := R^{\sigma-1} e^{AT_s}, \quad \sigma \in \{1, 2\}. \tag{5.17}$$

Proposition 5.2. *RCS (5.13) is asymptotically stable if and only if the switched system (5.15)–(5.17) is asymptotically stable.*

Proof: The proof is based on the fact that the state η_k of system (5.15)–(5.17) is simply the state of system (5.13) at kT_s, i.e., $\eta_k = \xi_k^+$. Thus, if (5.13) is asymptotically stable, so is system (5.15)–(5.17) and the necessity follows. Now suppose that the switched system (5.15)–(5.17) is asymptotically stable. Note that for any t, there is a k such that $t \in (kT_s, (k+1)T_s]$ and

$$\xi(t) = e^{A(t-kT_s)} \xi_k^+ = e^{A(t-kT_s)} \eta_k.$$

The boundedness of T_s implies that the matrix $e^{A(t-kT_s)}$ is bounded. It follows that (5.13) is stable and

$$\xi(t) \to 0, \quad \text{as } t \to \infty$$

since $\eta_k \to 0$ as $k \to \infty$. Thus (5.13) is asymptotically stable.

By Proposition 5.2, the following common Lyapunov-function-based result follows immediately [4, 5].

Proposition 5.3. *If there is a matrix $P > 0$ such that*

$$A_\sigma^T P A_\sigma - P < 0, \quad \forall \sigma \in \{1, 2\}, \tag{5.18}$$

then reset system (5.13) is asymptotically stable.

When a common Lyapunov function does not exist, we resort to other tools to check the stability. There are several tools, which have been developed for switched systems, that we can borrow to check the stability of RCSs. For simplicity, for any positive integer $m \ge 1$, denote

$$A_{j_m \cdots j_1} := A_{j_m} \cdots A_{j_1} \tag{5.19}$$

where $j_i = 1$ or 2 and A_1 and A_2 are defined in (5.17).

Proposition 5.4. *If there are symmetric matrices $P_{i_1 \cdots i_m}$, nonnegative numbers $\alpha^s_{i_1 \cdots i_m}$ and $\beta^s_{i_1 \cdots i_m j}$, $i_1, \ldots, i_m, j \in \{1, 2\}$; $s = 1, \ldots, m$, such that the following*

LMIs are satisfied

$$P_{i_1\cdots i_m} + \sum_{s=1}^{m} (-1)^{i_s} \alpha^s_{i_1\cdots i_m} A^T_{i_{s-1}\cdots i_1} Q A_{i_{s-1}\cdots i_1} > 0 \tag{5.20}$$

$$P_{i_1\cdots i_m} - A^T_{i_1} P_{i_2\cdots i_m j} A_{i_1} + \sum_{s=1}^{m} (-1)^{i_s} \beta^s_{i_1\cdots i_m} A^T_{i_{s-1}\cdots i_1} Q A_{i_{s-1}\cdots i_1} > 0 \tag{5.21}$$

$$i_1, \ldots, i_m, j = 1, 2, \tag{5.22}$$

where m is a positive integer and Q is defined in (5.14), then the RCS (5.13) is globally asymptotically stable.

Proof: The first step is to partition the state space $\mathbb{R}^{n_p+n_r}$ into 2^m disjoint subsets $\Omega_{j_1\cdots j_m}, j_i = 1$ or 2. We start with the case $m = 1$. Define

$$\Psi_0 := \{\eta \mid \eta^T Q \eta > 0\}.$$

Then

$$\mathbb{R}^{n_p+n_r} = \Omega_1 \cup \Omega_2$$

with $\Omega_1 := \Psi_0$, $\Omega_2 := \Psi_0^c$. Here, the superscript "c" denotes the complementary set. For the case of $m = 2$, we partition each Ω_i ($i = 1, 2$) obtained above to get a finer partition of $\mathbb{R}^{n_p+n_r}$ as follows. Define

$$\Psi_j = \{\eta \mid \eta^T A_j^T Q A_j \eta > 0\}, \quad j = 1, 2$$

and

$$\Omega_{j1} := \Omega_j \cap \Psi_j, \quad \Omega_{j2} := \Omega_j \cap \Psi_j^c, \quad j = 1, 2.$$

Thus we get four subsets $\Omega_{j_1 j_2}$, $j_1, j_2 = 1, 2$, which constitute a disjoint partition of $\mathbb{R}^{n_p+n_r}$. Generally, suppose that we have partitioned $\mathbb{R}^{n_p+n_r}$ into 2^{m-1} disjoint subsets $\Omega_{j_1 j_2\cdots j_{m-1}}$, $j_1, \ldots, j_{m-1} = 1$ or 2. We can similarly partition each $\Omega_{j_1 j_2\cdots j_{m-1}}$ into two disjoint subsets $\Omega_{j_1 j_2\cdots j_{m-1} j_m}$ ($j_m = 1, 2$) as follows:

$$\Psi_{j_1\cdots j_m} := \{\eta \mid \eta^T A^T_{j_m\cdots j_1} Q A_{j_m\cdots j_1} \eta > 0\},$$

$$\Omega_{j_1\cdots j_m} := \begin{cases} \Omega_{j_1\cdots j_{m-1}} \cap \Psi_{j_1\cdots j_{m-1}}, & j_m = 1 \\ \Omega_{j_1\cdots j_{m-1}} \cap \Psi^c_{j_1\cdots j_{m-1}}, & j_m = 2 \end{cases}$$

where $j_1 \cdots j_m = 1$ or 2 and the notation $A_{j_m\cdots j_1}$ is defined in (5.19). By doing this, the state space $\mathbb{R}^{n_p+n_r}$ is partitioned into 2^m disjoint subsets $\Omega_{j_1 j_2\cdots j_m}$. This partition has two properties which will be used later:

1. For any $\eta \in \Omega_{i_1\cdots i_m}$ and any $s = 1, 2, \ldots, m$, there holds
   $(-1)^{i_s} \eta^T A^T_{i_{s-1}\cdots i_1} Q A_{i_{s-1}\cdots i_1} \eta \leq 0$[1].
2. Suppose that $\{\eta_k\}$ is a solution to the switched system (5.15)–(5.17). Then $\eta_k \in \Omega_{i_1\cdots i_m}$ implies $\eta_{k+1} \in \Omega_{i_2\cdots i_m j}$ for some $j = 1$ or 2.

[1]When $s = 1$, $A_{i_{s-1}\cdots i_1} := I$.

The property (1) directly comes from the definition of $\Omega_{j_1j_2\cdots j_m}$ and $\Psi_{j_1j_2\cdots j_m}$. We prove property (2) as follows. Note that

$$\eta_k \in \Omega_{i_1\cdots i_m} \subseteq \Omega_{i_1\cdots i_{m-1}} \subseteq \cdots \subseteq \Omega_{i_1i_2} \subseteq \Omega_{i_1} \tag{5.23}$$

and by definition of $\Omega_{i_1\cdots i_s}$, (5.23) implies that, for any $s = 0, 1, 2, \ldots, m-1$,

$$\eta_k \in \begin{cases} \Psi_{i_1\cdots i_s}, & i_{s+1} = 1 \\ \Psi^c_{i_1\cdots i_s}, & i_{s+1} = 2. \end{cases} \tag{5.24}$$

(When $s = 0$, $\Psi_{i_1\cdots i_s} := \Psi_0$.) Letting $s = 0$ in (5.24) leads to $\eta_k^T Q\eta_k > 0$ if $i_1 = 1$ and $\eta_k^T Q\eta_k \leq 0$ if $i_1 = 2$. Thus $\sigma(\eta_k) = i_1$ and $\eta_{k+1} = A_{i_1}\eta_k$ by definition of $\sigma(\,\cdot\,)$ in (5.16). Letting $s = 1$ in (5.24) leads to $\eta_k \in \Psi_{i_1}$ if $i_2 = 1$ and $\eta_k \in \Psi^c_{i_1}$ if $i_2 = 2$. This fact together with $\eta_{k+1} = A_{i_1}\eta_k$ implies $\eta_{k+1}^T Q\eta_{k+1} > 0$ if $i_2 = 1$ and $\eta_{k+1}^T Q\eta_{k+1} \leq 0$ if $i_2 = 2$. Or equivalently, $\eta_{k+1} \in \Omega_{i_2}$.

By letting $s = 2, \ldots, m-1$ successively in (5.24) together with the definition of $\Psi_{i_1\cdots i_s}$ and the fact $\eta_{k+1} = A_{i_1}\eta_k$, there holds that

$$\eta_{k+1}^T A^T_{i_s\cdots i_2} Q A_{i_s\cdots i_2}\eta_{k+1} \begin{cases} > 0, & i_{s+1} = 1 \\ \leq 0, & i_{s+1} = 2 \end{cases}$$

This in turn implies that $\eta_{k+1} \in \Psi_{i_2\cdots i_s}$ if $i_{s+1} = 1$ and $\eta_{k+1} \in \Psi^c_{i_2\cdots i_s}$ if $i_{s+1} = 2$. By this fact and $\eta_{k+1} \in \Omega_{i_2}$, we can successively get

$$\eta_{k+1} \in \Omega_{i_2i_3}, \quad \eta_{k+1} \in \Omega_{i_2i_3i_4}, \quad \ldots, \quad \eta_{k+1} \in \Omega_{i_2i_3\ldots i_m}.$$

Thus property (2) is established by noting that

$$\Omega_{i_2i_3\cdots i_m} = \Omega_{i_2i_3\cdots i_m1} \cup \Omega_{i_2i_3\cdots i_m2}.$$

Now we are ready to define the Lyapunov function candidate used to prove the stability as

$$V(\eta) = \eta^T P_{i_1\cdots i_m}\eta, \quad \eta \in \Omega_{i_1\cdots i_m} \tag{5.25}$$

where $P_{i_1\cdots i_m}$ is the solution to LMIs (5.20) and (5.22). The condition (5.20) and property (1) imply that $V(\eta)$ is positive definite. In the following, we prove that the single-step forward difference $V(\eta_{k+1}) - V(\eta_k)$ is negative. The property (2) established above implies that we only need to check the case that $\eta_k \in \Omega_{i_1\cdots i_m}$ and $\eta_{k+1} \in \Omega_{i_2\cdots i_{m-1}j}$. Thus by (5.22)

$$\begin{aligned} &V(\eta_{k+1}) - V(\eta_k) \\ &= \eta_{k+1}^T P_{i_2\cdots i_mj}\eta_{k+1} - \eta_k^T P_{i_1\cdots i_m}\eta_k \\ &= \eta_k^T A_{i_1}^T P_{i_2\cdots i_mj} A_{i_1}\eta_k - \eta_k^T P_{i_1\cdots i_m}\eta_k \\ &< \sum_{s=1}^{m} (-1)^{i_s}\beta^s_{i_1\cdots i_m}\eta_k^T A^T_{i_{s-1}\cdots i_1} Q A_{i_{s-1}\cdots i_1}\eta_k \\ &\leq 0. \end{aligned}$$

Note that within each subset $\Omega_{i_1\cdots i_m}$, $V(\eta)$ and $V(\eta_{k+1}) - V(\eta_k)$ are all of quadratic form and the number of subsets is finite, it follows that there are $\mathcal{K}_\infty$ functions α_1, α_2, and α_3 such that for any $\eta \in \mathbb{R}^{n_p+n_r}$, there hold

$$\alpha_1(\|\eta\|) \le V(\eta) \le \alpha_2(\|\eta\|)$$

and

$$V(\eta_{k+1}) - V(\eta_k) \le -\alpha_3(\|\eta_k\|).$$

Thus system (5.15) is globally asymptotically stable [6].

The following corollary is a special case of Proposition 5.4 for $m = 1$.

Corollary 5.1. *If there are two symmetric matrices P_i, six nonnegative numbers α_i, β_{ij}, $i,j \in \{1,2\}$ such that*

$$P_i + (-1)^i \alpha_i Q > 0 \tag{5.26}$$

$$P_i - A_i^T P_j A_i + (-1)^i \beta_{ij} Q > 0 \tag{5.27}$$

$$\forall i,j = 1,2,$$

then the RCS (5.13) is globally asymptotically stable.

Remark 5.1. *For discrete-time discontinuous systems, discontinuous uniformly strict Lyapunov function candidates, which are adopted to prove Proposition 5.4, do not necessarily imply robustness with respect to disturbance. Additional conditions should be imposed on $P_{i_1\cdots i_m}$ to ensure the input-to-state stability of the disturbed switched system*

$$\eta_{k+1} = A_{\sigma(\eta_k)}\eta_k + w_k.$$

See Reference 6 for details.

5.3 A heuristic design method

In this section, we first investigate the role that the reset action plays in a well-designed RCS. Then we propose a guideline for designing a reset controller for SISO systems with good transient performance.

Consider the closed-loop system (5.9). If for any k, no reset occurs, then (5.9) degenerates to the baseline system

$$\begin{cases} \dot{x} = Ax + Kr \\ y = Cx. \end{cases} \tag{5.28}$$

On the other hand, if reset occurs at every sampling instant, then the evolutions of x_k^+ and y_k are governed by the discrete-time system

$$\begin{cases} \eta_{k+1} = M\eta_k + Nr \\ y_k = C\eta_k \end{cases} \tag{5.29}$$

where $\eta_k = x_k^+$ and

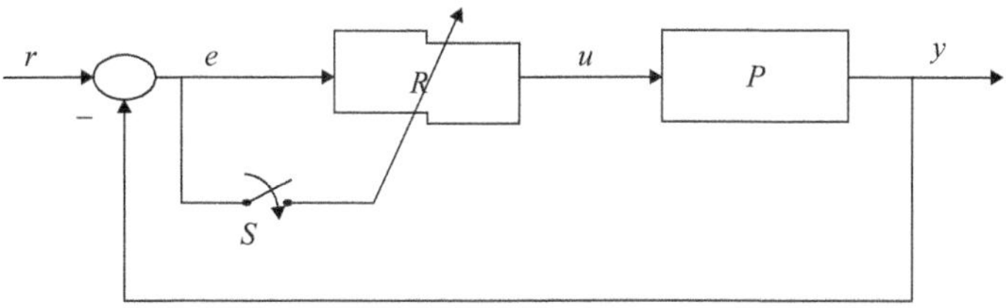

Figure 5.1 An RCS with discrete-time triggering condition

$$M = Re^{AT_s}, \tag{5.30}$$

$$N = L + R\left[\int_0^{T_s} e^{A(T_s-s)}ds\right]K. \tag{5.31}$$

We call system (5.29) the *reset mode*. The reset system (5.9) is actually a hybrid one combining the baseline system (5.28) with the reset mode (5.29). The triggering function $\phi(e_k, e_{k-1})$ governs when the reset mode (5.29) should come into play.

In order to understand more clearly the role that the reset mode plays, let us consider an example depicted in Figure 5.1. In this figure, P is the plant whose transfer function is

$$\frac{s+1}{s(s+0.2)}$$

and R is the reset controller with a baseline transfer function $\frac{1}{s+1}$. S is the sampler. This system with continuous triggering condition $e(t) = 0$ and zero reset matrix $R_r = 0$ has been discussed in Reference 7. Here, we consider the reset controller

$$\begin{cases} \dot{z} = -z + e, & t \neq kT_s \\ z_k^+ = z_k, & e_k e_{k-1} > 0 \\ z_k^+ = R_r z_k, & e_k e_{k-1} \leq 0 \\ u = z \end{cases} \tag{5.32}$$

with $e_k = e(kT_s)$ and R_r a scalar. The sampling period is chosen as $T_s = 0.1$. By using Corollary 5.1 and a simple search, one knows that this RCS is asymptotically stable for any $R_r \in [-1, 1]$. Figure 5.2 gives the step responses of the baseline system, the reset mode with $R_r = 0$ and the RCS with $R_r = 0$. From this figure, it is clear that the reset mode is much more damped than the baseline system. Combining these two leads to less overshoot with the same rise time. The role that the reset mode plays is to increase the damping and thus reduce the overshoot. Figure 5.3 gives the step responses of the reset mode for different R_r and shows that R_r has significant effect on the damping ratio of the reset mode. If we decrease R_r so that the reset mode is overdamped, less overshoot can be expected. This is confirmed in Figure 5.4 which gives the step responses of the overall RCS for different R_r.

By the above analysis, we summarize that a well-designed RCS can be viewed as a hybrid system obtained by combining an underdamped baseline system with an

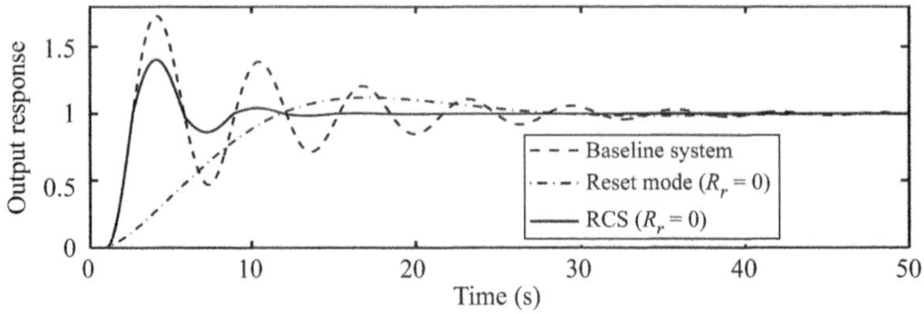

Figure 5.2 Step responses for the baseline system, reset mode, and the RCS with $R_r = 0$

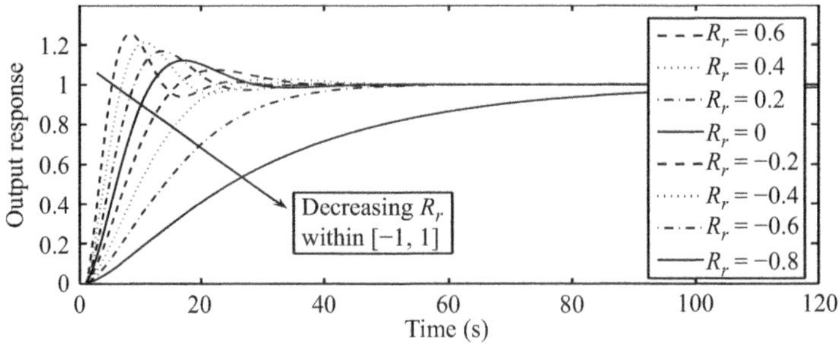

Figure 5.3 Step responses of the reset mode for different $R_r \in [-1, 1]$

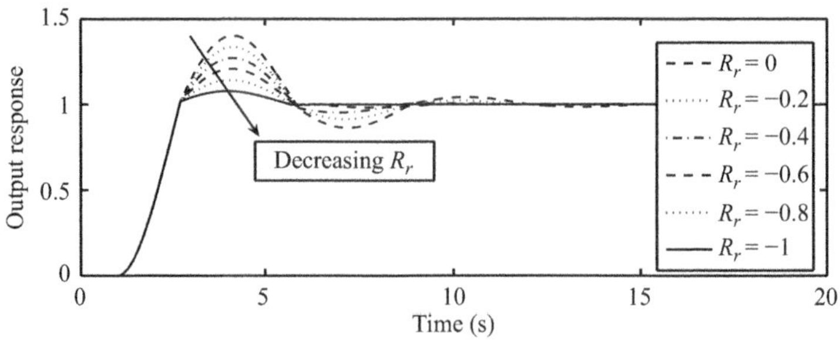

Figure 5.4 Step responses of the RCS for different $R_r \in [-1, 0]$

overdamped reset mode. The damping ratio can be changed by adjusting the reset matrix. This motivates the following guideline for control design.

Step 1. Design a baseline continuous controller

$$\begin{cases} \dot{z} = A_r z + B_r y + K_r r \\ u = F_r z + G_r y + H_r r \end{cases} \tag{5.33}$$

such that the output of the baseline system (5.28) tracks the reference input r asymptotically. This step is similar to the Horowitz's method. The principle is to meet other specifications without considering the overshoot constraint. When a baseline controller is designed, matrices A and K in (5.28) can be calculated through (5.10). Also, the steady-state value of $\mathbf{z}(r)$ can be calculated as a function of r. In general, there is an $n_r \times n_o$ matrix U such that

$$\mathbf{z}(r) = Ur.$$

Step 2. Adjust the matrices R_r and J_r so that the matrix $M = Re^{AT_s}$ has a large (s-plane equivalent) dominant damping ratio. Here, T_s is the pre-specified sampling period and the reset matrix R is given by (5.11). When J_r and R_r are determined, then by (5.12),

$$L_r = -J_r - (R_r - I)U.$$

Step 3. Tune the triggering function $\phi(e_k, e_{k-1})$ so that the baseline system and the reset mode cooperate well. For instance, suppose that $\phi(\alpha, \beta)$ is given in (5.5) and S in (5.8). When the sampling period T_s is fixed, there is only one parameter $\Delta > 0$ which needs to be tuned. By the definition of S in (5.8), Δ roughly characterizes how much time the resettings are ahead of the zero-crossings. This guides us to adjust Δ as follows: If the output response exhibits a large overshoot, increase Δ. On the contrary, if the system is overdamped, reduce Δ accordingly.

5.4 Application to track-seeking control of HDD systems

5.4.1 System description

Consider a HDD system modeled as

$$G(s) = G_u(s)G_n(s)$$

where $G_n(s)$ is the nominal model and $G_u(s)$ is the high-frequency model [8]. The state-space realization of the nominal model $G_n(s)$ is given by

$$\begin{cases} \dot{x}_p = A_p x_p + B_p u \\ y = C_p x_p \end{cases} \tag{5.34}$$

with

$$C_p = [1 \quad 0], \quad A_p = \begin{bmatrix} 0 & 1 \\ -a & -b \end{bmatrix}, \quad B_p = \begin{bmatrix} 0 \\ c \end{bmatrix},$$

$a = 1.4223 \times 10^5$, $b = 3.016 \times 10^2$, and $c = 1.134 \times 10^7$. $x_p = [x_{p1} \ \ x_{p2}]^T$ with x_{p1} (or y) the position of the *read/write* (R/W) head measured in micrometer (μm), x_{p2} the velocity of the R/W head in micrometer per second (μm/s)[2], and u the control input

[2]The actual output of the system (i.e., the position of the R/W head) should be $y_{\text{acutal}} = G_u(s)y$. Since our design is based on the nominal model, we identify y_{acutal} with y in the design stage. In the simulation, the high-frequency model will be included.

in volt (V). For any reference r, the steady-state values of x and u are, respectively, given by

$$\mathbf{x}_{p1}(r) = r, \quad \mathbf{x}_{p2}(r) = 0, \quad \mathbf{u}(r) = \frac{a}{c}r.$$

The high-frequency model is given by

$$G_u(s) = G_{u1}(s)G_{u2}(s)G_{u3}(s)$$

with

$$G_{u1}(s) = \frac{s^2 + 2711s + 2.047 \times 10^9}{s^2 + 3821s + 1.469 \times 10^9},$$

$$G_{u2}(s) = \frac{s^2 + 1.508 \times 10^4 s + 1.421 \times 10^9}{s^2 + 3382s + 1.144 \times 10^9},$$

$$G_{u3}(s) = \frac{s^2 + 1.407 \times 10^4 s + 1.238 \times 10^9}{s^2 + 6.256 \times 10^4 s + 1.996 \times 10^9}.$$

Our aim is to design a short-span-seeking controller with good transient performance.

5.4.2 Baseline controller design

We first design an observer-based baseline controller. A reduced order observer for x_2 of the form

$$\begin{cases} \dot{z}_1 = \lambda y + \tau z_1 + cu \\ \tilde{x}_{p2} = z_1 + \nu y \end{cases} \tag{5.35}$$

with

$$\nu = -b - \tau, \quad \lambda = -a - (b + \tau)\tau$$

and a full state controller

$$\begin{cases} \dot{z}_2 = e \\ u = \alpha e - \beta x_{p2} + \gamma z_2 \end{cases} \tag{5.36}$$

are designed. In the above equation, $e = r - y$ is the tracking error. By replacing x_{p2} in (5.36) by the estimated value $\tilde{x}_{p2}$, the overall baseline controller is then of the form (5.33) with $z = (z_1 \;\; z_2)^T$,

$$A_r = \begin{bmatrix} \tau - c\beta & c\gamma \\ 0 & 0 \end{bmatrix}, \quad B_r = \begin{bmatrix} \lambda - c\alpha - c\beta\nu \\ -1 \end{bmatrix}, \quad K_r = \begin{bmatrix} c\alpha \\ 1 \end{bmatrix}$$

and

$$F_r = [-\beta \;\; \gamma], \quad G_r = -(\alpha + \beta\nu), \quad H_r = \alpha.$$

We choose the observer pole as $\tau = -4 \times 10^3$. The baseline controller parameters are chosen as $\alpha = 2$, $\beta = 2 \times 10^{-3}$, and $\gamma = 3 \times 10^4$. Thus the baseline system is of the form (5.28) with $x = (x_{p1}\ x_{p2}\ z_1\ z_2)^T$ and

$$A = \begin{bmatrix} 0 & 1 & 0 & 0 \\ -1.0670 \times 10^8 & -301.6 & -22680 & 3.402 \times 10^{11} \\ -1.2150 \times 10^8 & 0 & -26680 & 3.402 \times 10^{11} \\ -1 & 0 & 0 & 0 \end{bmatrix},$$

$$K = \begin{bmatrix} 0 \\ 2.268 \times 10^7 \\ 2.668 \times 10^7 \\ 1 \end{bmatrix}, \quad C = \begin{bmatrix} 1 & 0 & 0 & 0 \end{bmatrix}.$$

The step response of the baseline system is given in Figure 5.5 (the dashed line) which shows that it is underdamped.

5.4.3 Reset mode

Only the controller state z_2 is to be reset and we use a reset law of the form

$$z_2^+ = R_r(z_2 - \mathbf{z}_2(r)) + \mathbf{z}_2(r), \quad [e_k\ \ e_{k-1}]S\begin{bmatrix} e_k \\ e_{k-1} \end{bmatrix} < 0$$

where S is of the form (5.8) with Δ to be determined later. $\mathbf{z}_2(r)$ is the steady-state value of z_2 given by

$$\mathbf{z}_2(r) = \frac{a}{c\gamma}r.$$

The reset law can also be represented in the form of

$$x^+ = Rx + Lr$$

with

$$R = \begin{bmatrix} 1 & 0 & 0 \\ 0 & 1 & 0 \\ 0 & 0 & R_r \end{bmatrix}, \quad L = \begin{bmatrix} 0 \\ 0 \\ 0 \\ \frac{(1-R_r)a}{c\gamma} \end{bmatrix}.$$

R_r is a scalar which is to be adjusted to change the damping ratio of the reset mode (5.29). We choose the sampling period $T_s = 2 \times 10^{-5}$ s. Thus for a given R_r, M and N can be calculated accordingly by (5.30) and (5.31). Figure 5.6 gives the step responses of the reset mode (5.29) for different R_r. It shows that R_r has significant effect on the damping ratio and the damping ratio increases with the decreasing of R_r. According to the guideline of design in the previous section, R_r should be small so that the reset mode is overdamped. For instance, $R_r \leq 0.8$.

5.4.4 Stability analysis

By Proposition 5.3 and some simple search, for any $R_r \in [0.3289,\ \ 1]$, A_1 and A_2 share a common Lyapunov function. One can easily see by using LMI tool that

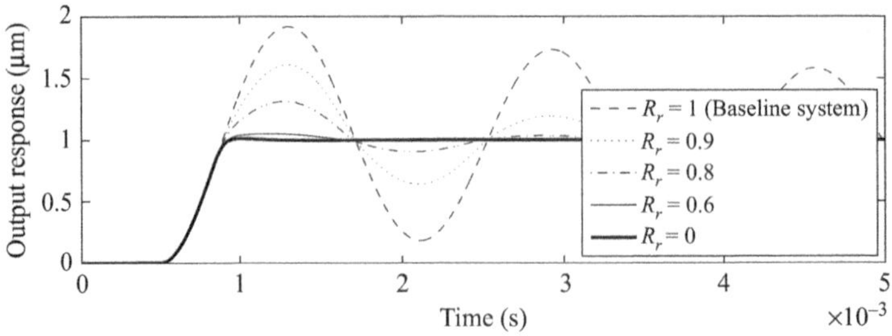

Figure 5.5 Step responses for different R_r ($\Delta = 4.5T_s$)

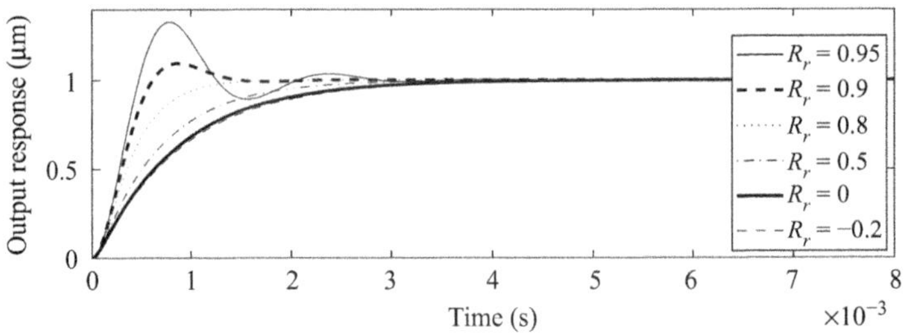

Figure 5.6 Step responses of the reset mode for different R_r

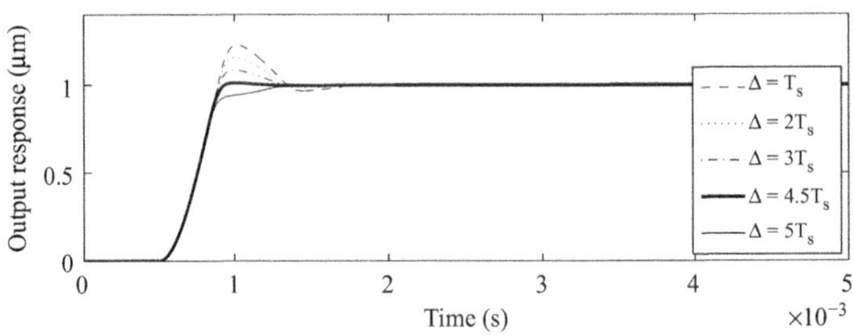

Figure 5.7 Step responses for different Δ ($R_r = 0$)

$R_r = 0.3289$ is very close to the marginal value for the existence of a common Lyapunov function. For $R_r < 0.3289$, we resort to Proposition 5.4 to check the stability. In this example, we assume that Q is of the form (5.14) with S given in (5.8). By some simple search, one sees that for any $R_r \in [-0.2, 0.4]$ and $\Delta \in [0, 5T_s]$, the conditions of Corollary 5.1 hold. Based on the analysis above, we can safely adjust R_r and Δ within the intervals $R_r \in [-0.2, 1]$ and $\Delta \in [0, 5T_s]$ without destroying the stability.

5.4.5 Simulation results

In the simulation, the control input is limited within $|u| \leq 3$ V. First of all, we choose $R_r = 0$. This choice is to assure a large damping ratio of the reset mode. Refer to

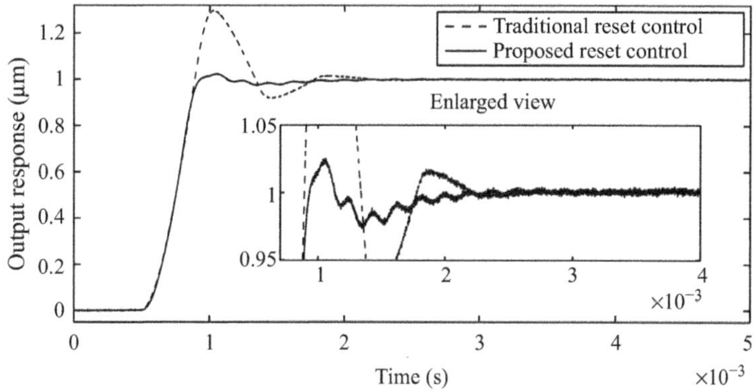

Figure 5.8 Comparison between traditional and proposed reset control

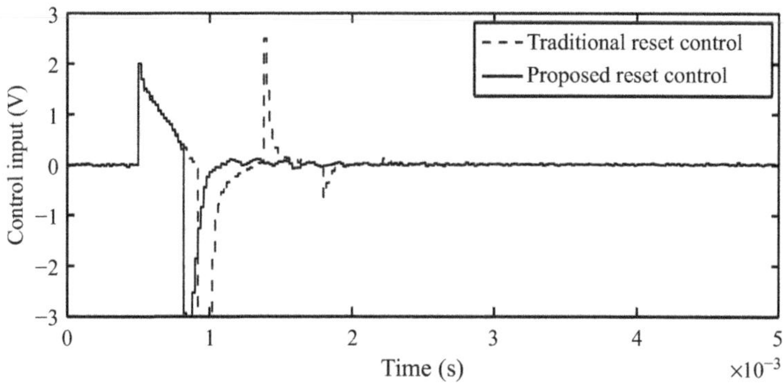

Figure 5.9 Control input

Figure 5.6. Also refer to Figure 5.5 where the step responses for different R_r with $\Delta = 4.5T_s$ are given. In Figure 5.7, the step responses for different Δ with $R_r = 0$ are given. From these two simulations, one can see that $R_r = 0$ and $\Delta = 4.5T_s$ provide perfect transient performance in this example.

For comparison, we also consider the traditional resetting, i.e., $R_r = 0$, under the triggering condition $e(t) = 0$. In simulation, the continuous triggering condition $e(t) = 0$ is replaced by its discrete-time version $e_k e_{k-1} < 0$. In this case, Q is of the form (5.14) with S given in (5.6). It is easy to check that this RCS is asymptotically stable by Corollary 5.1. In the following simulation, the high-frequency model is inserted and input disturbance, output disturbance, and measurement noise are added to simulate the real environment. In Figure 5.8, the step responses of the traditional RCS and the proposed RCS are given. Figure 5.9 shows the corresponding control efforts. This simulation shows that the proposed reset control achieves better transient performance and exhibits certain robustness with respect to model uncertainty, disturbance, and noise.

Notes

This chapter is based on Reference 9. There are many ways to define discrete-time triggering conditions and (5.1) is the simplest one which was first adopted in Reference 10. This chapter is also related to the idea of variable reset band which was first introduced into the reset control design in References 11 and 12. For the recent development of discrete-time reset control systems, we refer the reader to Reference 13.

Proposition 5.4 is based on the piecewise quadratic Lyapunov function technique [6, 14]. Some other tools for switched systems can be borrowed to obtain different stability criteria for RCSs and they include the multiple Lyapunov function technique [4, 15], dwell-time-based analysis [4, 16], etc. We also refer the reader to References 15, 14, 16, 4, 5, and 6, for stability of switched systems.

References

[1] J. Carrasco, A. Baños, and A. van der Schaft. A passivity-based approach to reset control systems stability. *Systems & Control Letters*, 59(1):18–24, 2010.

[2] W.M. Haddad, V.S. Chellaboina, and S.G. Nersesov. Impulsive and hybrid dynamical systems: stability, dissipativity, and control. In *Princeton Series in Applied Mathematics*, Princeton University Press, Princeton, New Jersey, 2006.

[3] D. Nešic, L. Zaccarian, and A.R. Teel. Stability properties of reset systems. *Automatica*, 44(8):2019–2026, 2008.

[4] D. Liberzon. *Switching in systems and control*. Birkhauser, Boston, MA, 2003.

[5] Z. Sun and S.S. Ge. Switched linear systems: control and design. In *Communications and Control Engineering*, Springer-Verlag London, Berlin, Heidelberg, 2005.

[6] M. Lazar, W.P.M.H. Heemels, and A.R. Teel. Lyapunov functions, stability and input-to-state stability subtleties for discrete-time discontinuous systems. *IEEE Transactions on Automatic Control*, 54(10):2421–2425, 2009.

[7] O. Beker, C.V. Hollot, Y. Chait, and H. Han. Fundamental properties of reset control systems. *Automatica*, 40(6):905–915, 2004.

[8] Y. Guo, Y. Wang, L. Xie, and J. Zheng. Stability analysis and design of reset systems: Theory and an application. *Automatica*, 45(2):492–497, 2009.

[9] Y. Guo, W. Gui, C. Yang, and L. Xie. Stability analysis and design of reset control systems with discrete-time triggering conditions. *Automatica*, 48(3):528–535, 2012.

[10] A. Banos, F. Perez, and J. Cervera. Discrete-time reset control applied to networked control systems. In *35th Annual Conference of IEEE Industrial Electronics Society*, November 3–5, 2009, IEEE, Porto, Portugal, pp. 2993–2998, 2009.

[11] A. Vidal and A. Baños. Reset compensation for temperature control: experimental application on heat exchangers. *Chemical Engineering Journal*, 159(1–3):170–181, 2010.

[12] F. Villaverde, A. Barreiro, J. Carrasco, and A. Banos. Reset control for passive bilateral teleoperation. *IEEE Transactions on Industrial Electronics*, 58(7):3037–3045, 2011.

[13] J. Carrasco and E.M. Navarro-López. Towards L_2-stability of discrete-time reset control systems via dissipativity theory. *Systems & Control Letters*, 62(6):525–530, 2013.

[14] L. Xie, S. Shishkin, and M. Fu. Piecewise Lyapunov functions for robust stability of linear time-varying systems. *Systems & Control Letters*, 31(3): 165–171, 1997.

[15] M.S. Branicky. Multiple Lyapunov functions and other analysis tools for switched and hybrid systems. *IEEE Transactions on Automatic Control*, 43(4):475–482, 1998.

[16] D. Liberzon and A.S. Morse. Basic problems in stability and design of switched systems. *IEEE Control Systems Magazine*, 19(5):59–70, 1999.

Chapter 6
Reset control systems with fixed reset instants

6.1 Stability analysis

6.1.1 Stability analysis through induced discrete systems

Consider the linear control system

$$\begin{cases} \dot{x} = Ax + Bu \\ y = Cx, \end{cases} \tag{6.1}$$

where $x \in \mathbb{R}^n$ is the state, $u \in \mathbb{R}^m$ the control input, $y \in \mathbb{R}^p$ the output of the system, and the reset controller

$$\begin{cases} \dot{z} = Dz + Ee, & t \neq t_k, \\ z(t^+) = F_k z + G_k e, & t = t_k, \\ u = Hz, \end{cases} \tag{6.2}$$

where $z \in \mathbb{R}^m$ and $e = r(t) - y$ with $r(t)$ being the reference signal. In this chapter, we assume that the reset time instants t_k are pre-specified. Combining (6.1) with (6.2), we get the closed-loop system

$$\begin{cases} \dot{\bar{x}} = \bar{A}\bar{x} + \bar{B}r, & t \neq t_k, \\ \Delta\bar{x} = \bar{R}_k\bar{x}, & t = t_k, \end{cases} \tag{6.3}$$

where $\bar{x} = (x, z)^T$, $\Delta\bar{x}(t) \triangleq x(t^+) - x(t)$, $x(t^+) \triangleq \lim_{s \to t^+} x(s)$, and

$$\bar{A} = \begin{pmatrix} A & BH \\ -EC & D \end{pmatrix}, \quad \bar{B} = \begin{pmatrix} 0 \\ E \end{pmatrix},$$

$$\bar{R}_k = \begin{pmatrix} 0 & 0 \\ -G_k C & F_k - I_m \end{pmatrix}$$

with I_m the identity matrix of dimension m.

In the following, we consider the stability of the general *impulsive linear system*

$$\begin{cases} \dot{x} = Ax, & t \neq t_k, \\ \Delta x = R_k x, & t = t_k. \end{cases} \tag{6.4}$$

We assume that the solution of system (6.4) is continuous from the left, thus

$$x(t_k^+) = (I + R_k)x(t_k),$$

where I is the identity matrix of the same dimension of x. It follows that

$$x(t_{k+1}^+) = (I + R_{k+1})e^{A(t_{k+1}-t_k)}x(t_k^+).$$

Denote $\eta_k = x(t_k^+)$, $\Delta t_k = t_k - t_{k-1}$, and

$$L_k = (I + R_k)e^{A\Delta t_k}.$$

Define $t_0 = 0$ and $R_0 = 0$, then we have

$$\eta_{k+1} = L_{k+1}\eta_k, \quad k = 0, 1, \ldots. \tag{6.5}$$

System (6.5) is called the *induced discrete system* of impulsive system (6.4).

Proposition 6.1. *If there exists a positive number $\Delta T > 0$ such that*

$$\Delta t_k = t_k - t_{k-1} < \Delta T \tag{6.6}$$

for all $k \in \mathbb{Z}^+$, then system (6.4) is (asymptotically) stable if and only if its induced discrete system (6.5) is (asymptotically) stable.

Proof: The necessity is obvious. We only prove the sufficiency. According to the fact that the solution of the base linear system

$$\dot{x} = Ax$$

depends continuously upon the initial condition, we have for any $\varepsilon > 0$, there exists a positive number $\delta_1 > 0$ such that

$$\|\zeta_0\| < \delta_1 \quad \Rightarrow \quad \|e^{As}\zeta_0\| < \varepsilon, \quad s \in [0, \Delta T]. \tag{6.7}$$

Assume that the induced discrete system (6.5) is stable, then for δ_1 selected above, there exists a positive number $\delta > 0$ such that

$$\|\eta_0\| < \delta \quad \Rightarrow \quad \|\eta_k\| < \delta_1, \quad k \in \mathbb{Z}^+.$$

Note that $\eta_0 = x_0$ and for any $t \in [0, +\infty)$, there is a nonnegative integer k such that $t \in (t_k, t_{k+1}]$, so

$$x(t) = e^{A(t-t_k)}x(t_k^+) = e^{A(t-t_k)}\eta_k.$$

Since $t - t_k \in [0, \Delta T]$, we have

$$\|x(t)\| = \|e^{A(t-t_k)}\eta_k\| < \varepsilon, \quad t \in [0, +\infty).$$

So system (6.4) is stable.

If the induced discrete system is asymptotically stable, we have $\lim_{k\to\infty} \eta_k = 0$. For any $\varepsilon > 0$, choose δ_1 such that

$$\|\zeta_0\| < \delta_1 \quad \Rightarrow \quad \|e^{As}\zeta_0\| < \varepsilon, \quad s \in [0, \Delta T]. \tag{6.8}$$

Thus for any η_0 ($=x_0$), there exists a $K(x_0) \in \mathbb{Z}^+$

$$\|\eta_k\| < \delta_1, \quad k \geq K(x_0).$$

Then for any $t > K(x_0)$, there exists a $k \geq K(x_0)$ such that $t \in [t_k, t_{k+1}]$ and $x(t) = e^{A(t-t_k)}\eta_k$, so we have

$$\|x(t)\| = \|e^{A(t-t_k)}\eta_k\| < \varepsilon$$

since $t - t_k \in [0, \Delta T]$. Thus $\lim_{t\to\infty} x(t) = 0$ and the system (6.4) is asymptotically stable.

From practical point of view, the base linear system is always designed to be stable, thus the bounded constraint on $\{\Delta t_k\}$ can be relaxed. We have the following result which can be proved similarly.

Proposition 6.2. *Assume the base linear system is stable, then system (6.4) is (asymptotically) stable if and only if its induced discrete system (6.5) is (asymptotically) stable.*

6.1.2 Lie-algebraic condition

In general, the induced discrete system of an impulsive system is time-varying, i.e., the structure matrix L_k depends upon k. It is well known if we can find a positive definite matrix P such that

$$L_k^T P L_k - P < 0, \quad k \in \mathbb{Z}^+,$$

then the induced discrete system is asymptotically stable. In fact, this is a problem of finding a common Lyapunov function. Note that common Lyapunov functions are also applied to investigate stability of switched systems under arbitrary switching laws.

Definition 6.1. *A vector space $\mathcal{L}$ over a field $\mathbb{F}$, with an operation $\mathcal{L} \times \mathcal{L} \to \mathcal{L}$ denoted by $(x, y) = [x, y]$ and called the bracket of x and y, is called a Lie algebra over $\mathbb{F}$ if the following axioms are satisfied*

1. *the bracket operation $[\cdot, \cdot]$ is bilinear.*
2. *$[x, x] = 0$ for all $x \in \mathcal{L}$.*
3. *$[x, [y, z]] + [y, [z, x]] + [z, [x, y]] = 0$ for any $x, y, z \in \mathcal{L}$.*

A subspace $\mathcal{K}$ of $\mathcal{L}$ is called a Lie subalgebra of $\mathcal{L}$ if

$$[x, y] \in \mathcal{K}, \quad \forall x, y \in \mathcal{K}.$$

A subspace I of a Lie algebra $\mathcal{L}$ is called a *ideal* of $\mathcal{L}$ if

$$[x, y] \in I, \quad \forall x \in \mathcal{L}, y \in I.$$

Define a sequence of ideals of $\mathcal{L}$ as $\mathcal{L}^{(0)} = \mathcal{L}$, $\mathcal{L}^{(1)} = [\mathcal{L}, \mathcal{L}]$, $\cdots$, $\mathcal{L}^{(i)} = [\mathcal{L}^{(i-1)}, \mathcal{L}^{(i-1)}], \ldots$. If there is an n such that $\mathcal{L}^{(n)} = 0$, then $\mathcal{L}$ is called a *solvable Lie algebra*.

The set of all $n \times n$ real matrices with the bracket being defined as

$$[A, B] = AB - BA$$

is a Lie algebra, denoted by $\mathfrak{g}(n,\mathbb{R})$. For any set of $n \times n$ matrices $\{A_p : p \in \mathscr{P}\}$, the smallest Lie subalgebra of $\mathfrak{g}(n,\mathbb{R})$ which contains $\{A_p : p \in \mathscr{P}\}$ is called the Lie algebra generated by $\{A_p : p \in \mathscr{P}\}$, denoted by $\{A_p : p \in \mathscr{P}\}_{LA}$.

Lemma 6.1. *[1] Assume that $\{A_p : p \in \mathscr{P}\}$ is a compact (with respect to the usual topology in $\mathbb{R}^{n\times n}$) set of Schur stable real matrices in the sense that*

$$|\lambda(A_p)| < 1, \quad \forall p \in \mathscr{P},$$

where the index set $\mathscr{P}$ is a subset of a finite-dimensional normal linear vector space. If the Lie algebra $\{A_p : p \in \mathscr{P}\}_{LA}$ generated by $\{A_p : p \in \mathscr{P}\}$ is solvable, then there exists a positive definite matrix P such that

$$A_p^T P A_p - P < 0, \quad p \in \mathscr{P}.$$

Remark 6.1. *Lie algebra $\{A_p : p \in \mathscr{P}\}_{LA}$ is solvable if and only if $\{A_p : p \in \mathscr{P}\}$ is simultaneously upper-triangularizable, i.e., there exists a nonsingular matrix Q such that $Q^{-1}A_pQ$ is of upper triangular form for any $p \in \mathscr{P}$. See Reference 1 for instance.*

Based on Proposition 6.1, we have the following.

Proposition 6.3. *Suppose that:*

1. *there exists a positive number ΔT such that $\Delta t_k < \Delta T$ for all $k \in \mathbb{Z}^+$;*
2. *$\{R_k\}$ is bounded and $\{A\} \cup \{R_k\}$ generates a solvable matrix Lie algebra;*
3. *there exists a positive number $\kappa > 0$ such that for any $k \in \mathbb{Z}^+$*

$$\ln|1 + \lambda(R_k)| + \Delta t_k \operatorname{Re}\lambda(A) \leq -\kappa < 0.$$

Then impulsive system (6.4) is asymptotically stable.

Proof: Denote

$$\mathbf{L} = \left\{(I + R_k)e^{A\Delta t_k} \mid k \in \mathbb{Z}^+\right\}. \tag{6.9}$$

It is obvious that $\mathbf{L}$ is a bounded set since $\{R_k\}$ and $\{\Delta t_k\}$ are bounded, hence its closure $\overline{\mathbf{L}}$ is compact in $\mathbb{R}^{q\times q}$, where q is the dimension of A. According to Proposition 6.1, we only need to prove that the induced discrete system is asymptotically stable, so it is sufficient to prove that $\overline{\mathbf{L}}$ has a common Lyapunov function since $\mathbf{L} \subset \overline{\mathbf{L}}$.

According to condition 2, $\{A\} \cup \{R_k\}$ is simultaneously upper-triangularizable. Thus there exists a nonsingular matrix Q such that for any $R \in \{A\} \cup \{R_k\}$, $\Gamma_R \triangleq Q^{-1}RQ$ is upper triangular. So for any $L \in \mathbf{L}$, we have

$$Q^{-1}LQ = Q^{-1}(I + R)e^{As}Q = (I + \Gamma_R)e^{\Gamma_{A}s} \triangleq \Gamma_L.$$

Γ_L is obviously upper triangular. Thus $\mathbf{L}$ is simultaneously upper-triangularizable and so is $\overline{\mathbf{L}}$. Hence $\overline{\mathbf{L}}$ generates a solvable Lie algebra. We have for any $k \in \mathbb{Z}^+$,

$$\lambda(L_k) = \lambda(\Gamma_{L_k}) \subset (1 + \lambda(\Gamma_{R_k}))e^{\lambda(\Gamma_A)\Delta t_k}.$$

Thus by condition 3,

$$\begin{aligned}\ln|\lambda(L_k)| &\subset \ln|1+\lambda(\Gamma_{R_k})| + \Delta t_k \operatorname{Re}(\lambda(\Gamma_A)) \\ &= \ln|1+\lambda(R_k)| + \Delta t_k \operatorname{Re}(\lambda(A)) \\ &\leq -\kappa < 0,\end{aligned}$$

hence we have $|\lambda(L_k)| \leq e^{-\kappa} < 1$. This leads to

$$|\lambda(L)| \leq e^{-\kappa} < 1, \quad L \in \overline{\mathbf{L}},$$

i.e., $\overline{\mathbf{L}}$ is a strictly Schur stable set. According to Lemma 6.1, there exists a positive definite matrix P such that

$$L^T PL - P < 0, \quad L \in \overline{\mathbf{L}}.$$

Since $\mathbf{L} \subset \overline{\mathbf{L}}$, we have

$$L_k^T PL_k - P < 0, \quad k \in \mathbb{Z}^+,$$

so the induced discrete system is asymptotically stable, and so is system (6.4).

Based on Lemma 6.1 and Proposition 6.2, we have the following result which can be proved similarly.

Proposition 6.4. *Suppose that:*

1. *the base linear system is stable;*
2. *$\{R_k\}$ is bounded and $\{A\} \cup \{R_k\}$ generates a solvable matrix Lie algebra;*
3. *there exists a positive number $\kappa > 0$ such that for any $k \in \mathbb{Z}^+$,*

$$\ln|1+\lambda(R_k)| + \Delta t_k \operatorname{Re}\lambda(A) \leq -\kappa < 0.$$

Then impulsive system (6.4) is asymptotically stable.

6.2 Moving horizon optimization

Consider the closed-loop *reset control system* (RCS)

$$\begin{cases} \dot{\bar{x}} = \bar{A}\bar{x} + \bar{B}r, & t \neq t_k, \\ z(t_k^+) = \rho_k(\bar{x}, r), & t = t_k, \\ y = \bar{C}\bar{x}, \end{cases} \tag{6.10}$$

where $\bar{x} = (x, z)^T$, $x \in \mathbb{R}^n$ is the state of the plant and $z \in \mathbb{R}^m$ is the state of the controller. $\rho_k(\bar{x}, r)$ is to be designed. Denote $e = r - y$, where the reference input r is assumed to be constant.

At every time instant t_k, consider a performance index of the form

$$J_k = \mathscr{J}_0(e(t_{k+N}), \dot{e}(t_{k+N})) + \int_{t_k}^{t_{k+N}} \mathscr{J}_1(e(s))ds \tag{6.11}$$

where $\mathscr{J}_0$ and $\mathscr{J}_1$ are positive functions and N is a pre-specified positive integer. The moving horizon optimization problem here is to find a sequence of after-reset states $\rho_k, \rho_{k+1}, \ldots, \rho_{t_{k+N-1}}$ to minimize J_k but only ρ_k is implemented at t_k. At the next time instant t_{k+1}, a sequence of after-reset states $\rho_{k+1}, \ldots, \rho_{t_{k+N}}$ are calculated to minimize J_{k+1} and only ρ_{k+1} is implemented. This process repeats over and over.

In this subsection, we only consider the case with $N = 1$. Thus our aim is to design ρ_k to minimize the following quadric performance index

$$J_k = \mathscr{J}_0(e(t_{k+1}), \dot{e}(t_{k+1})) + \int_{t_k}^{t_k+1} \mathscr{J}_1(e(s))ds, \tag{6.12}$$

$$\mathscr{J}_0 = e^T(t_{k+1})P_0 e(t_{k+1}) + \dot{e}^T(t_{k+1})Q_0\dot{e}(t_{k+1}),$$
$$\mathscr{J}_1 = e^T P_1 e,$$

at each reset time instant t_k, where P_0, P_1, and Q_0 are positive semi-definite matrices. For $N \geq 2$, the calculation is much more involved in which case we can use the method in Section 6.3 to calculate the optimal after-reset states.

Assumption 6.1. *Assume that for any $r \in \mathbb{R}^p$, there exists $\bar{\mathbf{x}}_r = (\mathbf{x}_r, \mathbf{z}_\mathbf{r})^T \in \mathbb{R}^{n+m}$ such that*

$$\begin{cases} \bar{A}\bar{\mathbf{x}}_r + \bar{B}r = 0, \\ \bar{C}\bar{\mathbf{x}}_r - r = 0. \end{cases} \tag{6.13}$$

Set

$$\begin{cases} \xi_x = x - \mathbf{x}_r, \\ \xi_z = z - \mathbf{z}_r, \end{cases} \tag{6.14}$$

we have

$$\begin{cases} \dot{\xi} = \bar{A}\xi, & t \neq t_k, \\ \xi_z(t_k^+) = \bar{\rho}_k(\xi, r), & t = t_k, \end{cases} \tag{6.15}$$

where $\xi = (\xi_x, \xi_z)^T$,

$$\bar{\rho}(\xi, r) = \rho(\xi + \bar{\mathbf{x}}_r, r) - \mathbf{z}_r.$$

Hence, we have

$$e = -\bar{C}\xi, \quad \dot{e} = -\bar{C}\bar{A}\xi.$$

Hence, the performance index function J_k can be rewritten as

$$J_k = \xi^T(t_{k+1})\bar{P}\xi(t_{k+1}) + \int_{t_k}^{t_{k+1}} \xi^T(s)\bar{Q}\xi(s)ds, \tag{6.16}$$

where

$$\bar{P} = \bar{C}^T P_0 \bar{C} + \bar{A}^T \bar{C}^T Q_0 \bar{C}\bar{A}, \tag{6.17}$$
$$\bar{Q} = \bar{C}^T P_1 \bar{C}. \tag{6.18}$$

When $\xi_x(t_k)$, r, and t_k, t_{k+1} are fixed, J_k is in fact a function of $\xi_z(t_k^+)$. In order to choose $\xi_z(t_k^+)$ such that J_k is minimized, we calculate $\frac{\partial J_k}{\partial \xi_z(t_k^+)}$ in the following. Note that

$$\xi(t) = e^{\bar{A}(t-t_k)}\xi(t_k^+), \quad t \in (t_k, t_{k+1}],$$

we have

$$\begin{aligned}\frac{\partial J_k}{\partial \xi(t_k^+)} &= \frac{\partial \xi(t_{k+1})}{\partial \xi(t_k^+)} \frac{\partial}{\partial \xi(t_{k+1})} \left(\xi^T(t_{k+1})\bar{P}\xi(t_{k+1})\right) \\ &\quad + \int_{t_k}^{t_{k+1}} \frac{\partial \xi(s)}{\partial \xi(t_k^+)} \frac{\partial}{\partial \xi(s)} \left(\xi^T(s)\bar{Q}\xi(s)\right) ds \\ &= 2e^{\bar{A}^T \Delta t_k}\bar{P}e^{\bar{A}\Delta t_k}\xi(t_k^+) + 2\int_{t_k}^{t_{k+1}} e^{\bar{A}^T(s-t_k)}\bar{Q}e^{\bar{A}(s-t_k)}\xi(t_k^+)ds \\ &= 2\Gamma_k \xi(t_k^+), \end{aligned} \tag{6.19}$$

where

$$\Gamma_k = e^{\bar{A}^T \Delta t_k}\bar{P}e^{\bar{A}\Delta t_k} + \int_0^{\Delta t_k} e^{\bar{A}^T s}\bar{Q}e^{\bar{A}s}ds. \tag{6.20}$$

Partition Γ_k as

$$\Gamma_k = \begin{pmatrix} \Gamma_k^{11} & \Gamma_k^{12} \\ \Gamma_k^{21} & \Gamma_k^{22} \end{pmatrix}$$

with $\Gamma_k^{12} = (\Gamma_k^{21})^T$. Thus we have

$$\begin{aligned}\frac{\partial J_k}{\partial \xi_z(t_k^+)} &= \frac{\partial \xi(t_k^+)}{\partial \xi_z(t_k^+)} \frac{\partial J_k}{\partial \xi(t_k^+)} \\ &= 2\begin{pmatrix} O & I \end{pmatrix} \begin{pmatrix} \Gamma_k^{11} & \Gamma_k^{12} \\ \Gamma_k^{21} & \Gamma_k^{22} \end{pmatrix} \xi(t_k^+) \\ &= 2(\Gamma_k^{21}\xi_x(t_k) + \Gamma_k^{22}\xi_z(t_k^+)) \end{aligned} \tag{6.21}$$

and the Hessian matrix of J_k with respect to $\xi_z(t_k^+)$ is

$$\mathrm{Hess}_{\xi_z(t_k^+)}(J_k) = 2\Gamma_k^{22}. \tag{6.22}$$

If Γ_k^{22} is positive definite, then by letting

$$\frac{\partial J_k}{\partial \xi_z(t_k^+)} = 0,$$

we obtain

$$\xi_z(t_k^+) = -(\Gamma_k^{22})^{-1}\Gamma_k^{21}\xi_x(t_k).$$

Thus the reset law which minimizes J_k is given by

$$\bar{\rho}_k(\xi, r) = -(\Gamma_k^{22})^{-1}\Gamma_k^{21}\xi_x(t_k). \tag{6.23}$$

Proposition 6.5. *Assume that Assumption 6.1 holds and $\Gamma_k^{22} > 0$. Then, the reset law which minimizes J_k is given by*

$$\rho_k(\bar{x}, r) = -(\Gamma_k^{22})^{-1}\Gamma_k^{21}(x - \mathbf{x}_r) + \mathbf{z}_r.$$

If we consider equidistant reset control, i.e., $\Delta t_k \equiv \Delta t$ is a constant, then Γ_k is a constant matrix which is independent of k and we denote it by Γ which is given by

$$\Gamma = e^{\bar{A}^T \Delta t} \bar{P} e^{\bar{A}\Delta t} + \int_0^{\Delta t} e^{\bar{A}^T s} \bar{Q} e^{\bar{A}s} ds.$$

Partition Γ similarly,

$$\Gamma = \begin{pmatrix} \Gamma^{11} & \Gamma^{12} \\ \Gamma^{21} & \Gamma^{22} \end{pmatrix}.$$

According to the analysis above along with Proposition 6.1, we have the following.

Proposition 6.6. *Suppose that $\Delta t_k = \Delta t$ is a constant and $\Gamma^{22} > 0$. Then, the reset law which minimizes J_k is independent of k and is given by*

$$\rho_k(\bar{x}, r) = -(\Gamma^{22})^{-1}\Gamma^{21}(x - \mathbf{x}_r) + \mathbf{z}_r.$$

In addition, under this reset law, the closed-loop system with $r = 0$ is asymptotically stable if and only if

$$\left| \lambda \left(\begin{pmatrix} I & O \\ -(\Gamma^{22})^{-1}\Gamma^{21} & O \end{pmatrix} e^{\bar{A}\Delta t} \right) \right| < 1. \tag{6.24}$$

6.2.1 Trade-off between stability and other performances

In the design of reset law, we have not taken into account stability. Thus under reset law (6.23), the stability of the closed-loop system may be destroyed. In this case, some trade-off should be made between the stability and performance characterized by J_k. One way to do this is to use the following reset law

$$\bar{\rho}_k(\xi, r, \mu) = -\mu(\Gamma_k^{22})^{-1}\Gamma_k^{21}\xi_x(t_k) + (1 - \mu)\xi_z(t_k), \quad \mu \in [0, 1]. \tag{6.25}$$

Suppose that the base linear system (i.e., $\mu = 0$) is asymptotically stable. When μ varies from 1 to 0, the closed-loop system changes from being unstable to being asymptotically stable.

On the other hand, according to (6.21), we have[1]

$$\begin{aligned}\frac{\partial J_k}{\partial \mu} &= \frac{\partial \xi_z(t_k^+)}{\partial \mu} \frac{\partial J_k}{\partial \xi_z(t_k^+)} \\ &= -2(1 - \mu)\left[(\Gamma_k^{22})^{-1}\Gamma_k^{21}\xi_x(t_k) + \xi_z(t_k)\right]^T \Gamma_k^{22}\left[(\Gamma_k^{22})^{-1}\Gamma_k^{21}\xi_x(t_k) + \xi_z(t_k)\right] \\ &\leq 0, \quad |\mu| \leq 1. \end{aligned} \tag{6.26}$$

Thus $J_k(\rho_k(\xi, r, \mu))$ is monotonously decreasing when μ varies from 0 to 1. We can always choose $\mu \in (0, 1)$ such that the closed-loop system is asymptotically stable and at the same time,

$$J_k(\rho_k(\xi, r, 1)) < J_k(\rho_k(\xi, r, \mu)) < J_k(\rho_k(\xi, r, 0)).$$

[1]The calculation of (6.26) is based on (6.21), (6.25), and the fact that $\xi_z(t_k^+) = \bar{\rho}_k(\xi, r, \mu)$.

The above inequality indicates that the performance index of the resulting RCS is always less than a system without resetting, though the minimal index may not be achieved.

By Proposition 6.1, one sees that for equidistant reset control, the overall system under reset law (6.25) is asymptotically stable if and only if

$$\left|\lambda\left(\begin{pmatrix} I & O \\ -\mu(\Gamma^{22})^{-1}\Gamma^{21} & (1-\mu)I \end{pmatrix} e^{\bar{A}\Delta t}\right)\right| < 1. \tag{6.27}$$

6.2.2 Observer-based reset control

In this section, we consider observer-based RCSs and show that reset controller and observer can be designed separately. For convenience, we assume that $r=0$ in this section. Consider RCS (6.1) and reset controller

$$\begin{cases} \dot{z} = Dz + Ex, & t \neq t_k, \\ \Delta z = G_k x + F_k z, & t = t_k, \\ u = Nx + Hz. \end{cases} \tag{6.28}$$

Thus the closed-loop system is given by

$$\begin{cases} \dot{\bar{x}} = \bar{A}\bar{x}, & t \neq t_k, \\ \Delta\bar{x} = \bar{R}_k\bar{x}, & t = t_k, \end{cases} \tag{6.29}$$

where

$$\bar{A} = \begin{pmatrix} A+BN & BH \\ E & D \end{pmatrix}, \quad \bar{R}_k = \begin{pmatrix} O & O \\ G_k & F_k \end{pmatrix}.$$

The induced discrete system is then of the form

$$\eta_{k+1} = L_{k+1}\eta_k, \tag{6.30}$$

where

$$L_k = (I + \bar{R}_k)e^{\bar{A}\Delta t_k}.$$

When the state of the plant is unmeasurable, x in the controller should be replaced by its estimated value $\tilde{x}$. We assume that (C, A) is observable and consider the following observer

$$\dot{\tilde{x}} = A\tilde{x} + Bu + K(y - C\tilde{x}), \tag{6.31}$$

where K is a matrix such that $A - KC$ is Hurwitz. Define $e_x = x - \tilde{x}$ and $\zeta = (\tilde{x}, z)^T$, then the overall closed-loop system is

$$\begin{cases} \begin{pmatrix} \dot{\zeta} \\ \dot{e}_x \end{pmatrix} = \begin{pmatrix} \bar{A} & * \\ O & A-KC \end{pmatrix}\begin{pmatrix} \zeta \\ e_x \end{pmatrix}, & t \neq t_k, \\ \Delta\zeta = \bar{R}_k\zeta, & t = t_k. \end{cases} \tag{6.32}$$

Denote $\bar{L}_k = e^{(A-KC)\Delta t_k}$. Then the induced discrete system of (6.32) is

$$\zeta_{k+1} = \tilde{L}_{k+1}\zeta_k, \tag{6.33}$$

where

$$\tilde{L}_k = \begin{pmatrix} L_k & H_k \\ O & \bar{L}_k \end{pmatrix}$$

and H_k depends on Δt_k. We have the following result.

Proposition 6.7. *Assume:*

1. *(C, A) is observable, and suppose K is a matrix such that $A - KC$ is stable;*
2. *controller (6.28) asymptotically stabilizes system (6.1) with a common Lyapunov function $P > 0$ satisfying*

   $$L_k^T P L_k - P < -Q, \quad k = 1, 2, \ldots,$$

   where $Q > 0$;
3. *$\{G_k\}$ and $\{F_k\}$ are bounded and*

   $$0 < \delta t \leq \Delta t_k \leq \Delta T < +\infty.$$

Then the closed-loop system (6.32) is asymptotically stable.

Proof: According to the boundedness of $\{G_k\}$, $\{F_k\}$, and $\{\Delta t_k\}$, we know that $\{\tilde{L}_k\}$ is bounded. Thus, $\overline{\{\tilde{L}_k\}}$ is a compact set and all of its elements are of the same block upper-triangular form. According to the stability of $A - KC$, $\{\bar{L}_k\}$ share a common Lyapunov function. Hence, the result is proved.

When reset controller is equidistant and $G_k = G_0$ and $F_k = F_0$ are constant matrices, some conditions of Proposition 6.7 can be relaxed. We have the following immediately.

Proposition 6.8. *Assume:*

1. *(C, A) is observable, and suppose K is a matrix such that $A - KC$ is stable;*
2. *$\Delta t_k \equiv \Delta T$, $G_k = G_0$, and $F_k = F_0$ do not rely on k.*

If controller (6.28) (asymptotically) stabilizes system (6.1), then (6.28) with x replaced by $\tilde{x}$ also (asymptotically) stabilizes system (6.1).

6.3 Optimal reset law design

Consider the plant (6.1) and a general reset controller of the form

$$\begin{cases} \dot{z} = Dz + Ee, & t \neq t_k, \\ z(t^+) = \rho_k(x, z, r), & t = t_k, \\ u = Gx + Hz + Me, & \end{cases} \tag{6.34}$$

where $z \in \mathbb{R}^q$ is the state of the controller and r is the reference signal which is assumed to be constant. D, E, G, H, and M are all constant matrices with compatible dimensions. $e = r - y$ is the tracking error. t_ks are the pre-specified reset time instants. $z(t^+) := \lim_{s \to t+0} z(s)$. $\rho_k = \rho_k(x, z, r)$ is the reset value at time instant $t = t_k$ which is to be designed and is called the *reset law*. Combining system (6.1) with reset controller (6.34) yields

$$\begin{cases} \dot{\bar{x}} = \bar{A}\bar{x} + \bar{B}r, & t \neq t_k, \\ z(t^+) = \rho_k(\bar{x}, r), & t = t_k, \\ y = \bar{C}\bar{x} \end{cases} \tag{6.35}$$

where $\bar{x} = (x^T, z^T)^T$, $\bar{C} = (C, \ \ 0_{p\times q})$, and

$$\bar{A} = \begin{pmatrix} A_c & BH \\ -EC & D \end{pmatrix}, \quad \bar{B} = \begin{pmatrix} BM \\ E \end{pmatrix},$$

$$A_c = A + BG - BMC.$$

Define the reset period as $\Delta t_k := t_{k+1} - t_k$. Assume that Assumption 6.1 holds. The steady-state control input $\mathbf{u}_r$ is given by

$$\mathbf{u}_r = G\mathbf{x}_r + H\mathbf{z}_r.$$

For each reset time interval $(t_k, t_{k+1}]$, we define a cost function J_k as

$$J_k = \int_{t_k}^{t_{k+1}} \left[e^T(s)Q_k e(s) + (u(s) - \mathbf{u}_r)^T R_k (u(s) - \mathbf{u}_r) \right] ds.$$

For the finite horizon case, define the cost $L(t_N)$ regarding the final state as

$$L(t_N) = e^T(t_N)Pe(t_N).$$

Here Q_k, R_k, and P are positive semi-definite matrices. The *optimal reset law* (ORL) design problems considered in this chapter are formulated as follows.

Problem 1 (Finite horizon case): Consider a finite time interval $[0, T]$ with reset time instants

$$0 = t_0 < t_1 < \cdots < t_N = T.$$

Design reset values $\rho_j, j = 0, 1, \ldots, N-1$, to minimize the cost function $J(t_N)$, where

$$J(t_N) := L(t_N) + \sum_{i=0}^{N-1} J_i. \tag{6.36}$$

Problem 2 (Infinite horizon case): Suppose that

$$\lim_{N\to\infty} t_N = \infty.$$

Design $\rho_j, j = 0, 1, \ldots,$ such that the resulting system is asymptotically stable and at the same time, the cost function $J(\infty)$ is minimized, where

$$J(\infty) := \sum_{i=0}^{\infty} J_i. \tag{6.37}$$

6.3.1 Equivalence between ORL and LQR

In this section, we turn to solve the ORL problems proposed in the previous section. The main idea is to transfer these problems to *linear quadratic regulation* (LQR) problems.

We first perform a coordinate transformation as

$$\begin{cases} \xi_x = x - \mathbf{x}_r, \\ \xi_z = z - \mathbf{z}_r, \end{cases} \tag{6.38}$$

where the steady-state values $\mathbf{x}_r$ and $\mathbf{z}_r$ are defined in Assumption 6.1. Thus (6.35) becomes

$$\begin{cases} \dot{\xi} = \bar{A}\xi, & t \neq t_k, \\ \xi_z(t_k^+) = \bar{\rho}_k, & t = t_k, \end{cases} \tag{6.39}$$

where $\xi = (\xi_x, \xi_z)^T$ and

$$\bar{\rho}_k = \rho_k - \mathbf{z}_r. \tag{6.40}$$

Hence we have

$$\begin{aligned} e &= -C\xi_x, \\ u &= (G - MC)\xi_x + H\xi_z + \mathbf{u}_r, \\ \xi(t_{k+1}) &= e^{\bar{A}\Delta t_k}\xi(t_k^+). \end{aligned}$$

Thus

$$\begin{aligned} J_k &= \int_{t_k}^{t_{k+1}} \big[\xi_x^T(s)C^T Q_k C\xi_x(s) + ((G - MC)\xi_x + H\xi_z)^T \\ &\qquad \times R_k((G - MC)\xi_x + H\xi_z)\big]\, ds \\ &= \xi^T(t_k^+)\Theta_k\xi(t_k^+) \end{aligned} \tag{6.41}$$

where

$$\begin{aligned} \Theta_k &:= \int_0^{\Delta t_k} e^{\bar{A}^T s}\Omega_k e^{\bar{A}s} ds, \\ \Omega_k &= \begin{pmatrix} \Omega_k^{11} & \Omega_k^{12} \\ \Omega_k^{12T} & \Omega_k^{22} \end{pmatrix}, \\ \Omega_k^{11} &= C^T Q_k C + (G - MC)^T R_k (G - MC), \\ \Omega_k^{12} &= (G - MC)^T R_k H, \\ \Omega_k^{22} &= H^T R_k H, \end{aligned}$$

and

$$\begin{aligned} L(t_N) &= \xi^T(t_{N-1}^+)e^{\bar{A}^T \Delta t_{N-1}}\bar{C}^T P\bar{C}e^{\bar{A}\Delta t_{N-1}}\xi(t_{N-1}^+) \\ &= \xi^T(t_{N-1}^+)\Phi_{N-1}\xi(t_{N-1}^+) \end{aligned} \tag{6.42}$$

where

$$\Phi_{N-1} := e^{\bar{A}^T \Delta t_{N-1}}\bar{C}^T P\bar{C}e^{\bar{A}\Delta t_{N-1}}.$$

Define

$$\bar{\Theta}_i = \begin{cases} \Theta_i, & i = 0, 1, \ldots, N-2, \\ \Theta_{N-1} + \Phi_{N-1}, & i = N-1. \end{cases}$$

Then we have

$$J(t_N) = \sum_{i=0}^{N-1} \xi^T(t_k^+)\bar{\Theta}_i \xi(t_k^+),$$
$$J(\infty) = \sum_{i=0}^{\infty} \xi^T(t_k^+)\Theta_i \xi(t_k^+).$$

On the other hand, from system (6.39), there holds

$$\begin{pmatrix} \xi_x(t_{k+1}) \\ \xi_z(t_{k+1}) \end{pmatrix} = e^{\bar{A}\Delta t_k} \begin{pmatrix} \xi_x(t_k) \\ \bar{\rho}_k \end{pmatrix}. \tag{6.43}$$

Partition $e^{\bar{A}\Delta t_k}$ as

$$e^{\bar{A}\Delta t_k} = \begin{pmatrix} \Gamma_A(k) & \Gamma_B(k) \\ * & * \end{pmatrix}, \tag{6.44}$$

then we obtain

$$\xi_x(t_{k+1}) = \Gamma_A(k)\xi_x(t_k) + \Gamma_B(k)\bar{\rho}_k. \tag{6.45}$$

System (6.45) is simply a time-varying linear discrete-time system with state variable $\xi_x(t_k)$ and control input $\bar{\rho}_k$.

Consider the following two LQR problems of system (6.45).

Problem A: (Finite horizon case) Design a control sequence $\bar{\rho}_k$, $k = 0, 1, \ldots, N-1$, for system (6.45) to minimize the quadratic performance index

$$J(t_N) = \sum_{k=0}^{N-1} \begin{pmatrix} \xi_x^T(t_k) & \bar{\rho}_k^T \end{pmatrix} \bar{\Theta}_k \begin{pmatrix} \xi_x(t_k) \\ \bar{\rho}_k \end{pmatrix}.$$

Problem B: (Infinite horizon case) Design a control sequence $\bar{\rho}_k$, $k = 0, 1, \ldots,$ for system (6.45) such that the resulting system is asymptotically stable and the following quadratic performance index is minimized,

$$J(\infty) = \sum_{k=0}^{\infty} \begin{pmatrix} \xi_x^T(t_k) & \bar{\rho}_k^T \end{pmatrix} \Theta_k \begin{pmatrix} \xi_x(t_k) \\ \bar{\rho}_k \end{pmatrix}.$$

For the finite horizon case, the following result is obvious.

Proposition 6.9. *The finite horizon ORL Problem 1 is equivalent to the LQR Problem A.*

For the infinite horizon case, we have the following.

Proposition 6.10. *Suppose that there exist positive numbers $\Delta T > 0$ and $\varepsilon > 0$ such that*

1. $\Delta t_k < \Delta T, \quad \forall k,$
2. $\lambda_{\min}(\Gamma_B^T(k)\Gamma_B(k)) \geq \varepsilon > 0, \quad \forall k.$

Then the infinite horizon ORL Problem 2 is equivalent to the LQR Problem B.

Proof: We only need to prove that under the conditions (1) and (2), system (6.45) is asymptotically stable if and only if system (6.39) is asymptotically stable. The *if* part is obvious. In the following, we prove the *only if* part. Assume that system (6.45) is asymptotically stable. In this case, we have

$$\lim_{k\to\infty} \xi_x(t_k) = 0,$$
$$\lim_{k\to\infty} \Gamma_B(k)\bar{\rho}_k = 0.$$

According to condition (2), we have

$$\lim_{k\to\infty} \bar{\rho}_k = 0.$$

Note that $\xi_z(t_k^+) = \bar{\rho}_k$ and $\{\Delta t_k\}$ is bounded, we have

$$\lim_{t\to\infty} \xi(t) = 0$$

and thus system (6.39) is asymptotically stable.

Remark 6.2. *For the infinite horizon case, we usually assume that the reset period and the matrices in the performance index are constant. In this case, $\Gamma_A(k)$ and $\Gamma_B(k)$ are also constant and the conditions of Proposition 6.10 hold naturally provided that Γ_B is of full rank. Actually, we can always make Γ_B be of full rank. Otherwise, system (6.45) would have redundant inputs in $\bar{\rho}_k$. In this case, we can put the redundant components of $\bar{\rho}_k$ and $\xi_z(t_k)$ into $\xi_x(t_k)$ by some rearrangements and transformations (refer to (6.43)) and then repartition $e^{\bar{A}\Delta t}$.*

Proposition 6.11. *Suppose that $(\Gamma_A(k), \Gamma_B(k))$ is controllable and there exist positive numbers $\delta t > 0$, $\Delta T > 0$, and $\varepsilon > 0$ such that:*

1. $\delta t \leq \Delta t_k \leq \Delta T, \quad \forall k,$
2. $\lambda_{\min}(\Omega_k) \geq \varepsilon > 0, \quad \forall k.$

Then the infinite horizon ORL Problem 2 is equivalent to the LQR Problem B.

Proof: Similarly, we only need to prove that the asymptotic stability of system (6.45) implies the asymptotic stability of system (6.39). According to the conditions and the definition of Θ_k, there is a positive number $\varepsilon' > 0$ such that

$$\lambda_{\min}(\Theta_k) \geq \varepsilon' > 0, \quad \forall k. \tag{6.46}$$

Suppose that $\{\bar{\rho}_k^*\}$ is the solution of the LQR Problem B. By controllability, the minimum value of $J(\infty)$ is finite and thus

$$\lim_{k\to\infty} \begin{pmatrix} \xi_x^T(t_k) & (\bar{\rho}_k^*)^T \end{pmatrix} \Theta_k \begin{pmatrix} \xi_x(t_k) \\ \bar{\rho}_k^* \end{pmatrix} = 0.$$

According to condition (6.46), we have

$$\lim_{k\to\infty} \xi(t_k^+) = \lim_{k\to\infty} \begin{pmatrix} \xi_x(t_k) \\ \bar{\rho}_k^* \end{pmatrix} = 0.$$

Since $\{\Delta t_k\}$ is bounded, we have $\lim_{t\to\infty} \xi(t) = 0$ and thus system (6.39) is asymptotically stable.

6.3.2 Solutions to ORL problems

According to the propositions in Section 6.3.1, in order to solve the ORL problems, we only need to solve the corresponding LQR problems. Suppose that we have obtained the optimal solution for the LQR problem of system (6.45),

$$\bar{\rho}_k^* = \bar{\rho}_k^*(\xi_x(t_k), \xi_z(t_k)),$$

then by (6.40), the solution for the corresponding ORL problem is given by

$$\rho_k^* = \bar{\rho}_k^*(x(t_k) - \mathbf{x}_r, z(t_k) - \mathbf{z}_r) + \mathbf{z}_r.$$

Problems A and B are standard LQR problems and thus we can obtain their optimal solutions directly by solving some Riccati equations. Partition Θ_k and Φ_{N-1} as

$$\Theta_k = \begin{pmatrix} \bar{Q}_k & \bar{T}_k \\ \bar{T}_k^T & \bar{R}_k \end{pmatrix}, \quad \Phi_{N-1} = \begin{pmatrix} \Phi_{11} & \Phi_{12} \\ \Phi_{12}^T & \Phi_{22} \end{pmatrix}.$$

Note that the weighting matrices Θ_k and Φ_{N-1} are not block diagonal. We can transfer this kind of the LQR problems to the block diagonal case through some transformations. The following result follows immediately.

Proposition 6.12. *Suppose that $\bar{\Theta}_k$ is positive definite, then the ORL which minimizes $J(N)$ is given by*

$$\rho_k^* = -K_k(x(t_k) - \mathbf{x}_r) + \mathbf{z}_r, \quad k = 0, 1, \ldots, N-1$$

where K_k and S_k are determined by

$$K_k = (\Gamma_B^T(k)S_{k+1}\Gamma_B(k) + \bar{R}_k)^{-1}(\Gamma_B^T(k)S_{k+1}\Gamma_A(k) + \bar{T}_k^T), \tag{6.47}$$
$$k = 0, \ldots, N-2$$

with

$$K_{N-1} = (\bar{R}_{N-1} + \Phi_{22})^{-1}(\bar{T}_{N-1} + \Phi_{12})^T$$

and

$$S_k = \Gamma_A^T(k)S_{k+1}\Gamma_A(k) + \bar{Q}_k - K_k^T(\Gamma_B^T(k)S_{k+1}\Gamma_B(k) + \bar{R}_k)K_k, \quad (6.48)$$
$$k = 0, \ldots, N-2$$

with

$$S_{N-1} = -K_{N-1}^T(\bar{R}_{N-1} + \Phi_{22})K_{N-1} + (\bar{Q}_{N-1} + \Phi_{11}).$$

Furthermore, the minimum of $J(t_N)$ is given by

$$J^*(t_N) = (x(0) - \mathbf{x}_r)^T S_0(x(0) - \mathbf{x}_r).$$

Remark 6.3. *The conditions (6.47) and (6.48) in Proposition 6.12 can also be given by the Riccati equation*

$$\begin{aligned} S_k &= \tilde{Q}_k + \bar{\Gamma}_A^T(k)S_{k+1}\bar{\Gamma}_A(k) - \bar{\Gamma}_A^T(k)S_{k+1}\Gamma_B(k) \\ &\quad \times(\Gamma_B^T(k)S_{k+1}\Gamma_B(k) + \bar{R}_k)^{-1}\Gamma_B^T(k)S_{k+1}\bar{\Gamma}_A(k), \\ k &= 0, 1, \ldots, N-2, \end{aligned}$$

with boundary condition

$$S_{N-1} = (\bar{Q}_{N-1} + \Phi_{11}) - (\bar{T}_{N-1} + \Phi_{12})(\bar{R}_{N-1} + \Phi_{22})^{-1}(\bar{T}_{N-1} + \Phi_{12})^T$$

where

$$\bar{\Gamma}_A(k) = \Gamma_A(k) - \Gamma_B(k)\bar{R}_k^{-1}\bar{T}_k^T, k = 0, 1, \ldots, N-2,$$
$$\tilde{Q}_k = \bar{Q}_k - \bar{T}_k\bar{R}_k^{-1}\bar{T}^T, \quad k = 0, 1, \ldots, N-2.$$

For Problem 2, it can be considered as the special case of Problem 1 when N tends to infinity. In the following, we assume that Θ_k is constant, denoted by Θ and that the reset time instants are equidistant, so $\Gamma_A(k)$ and $\Gamma_B(k)$ are constant, denoted by Γ_A and Γ_B, respectively. Partition Θ as

$$\Theta = \begin{pmatrix} \bar{Q} & \bar{T} \\ \bar{T}^T & \bar{R} \end{pmatrix}. \quad (6.49)$$

Define

$$\bar{\Gamma}_A = \Gamma_A - \Gamma_B\bar{R}^{-1}\bar{T}^T \quad (6.50)$$
$$\tilde{Q} = \bar{Q} - \bar{T}\bar{R}^{-1}\bar{T}^T. \quad (6.51)$$

We have the following result.

Proposition 6.13. *Assume that (Γ_A, Γ_B) is stabilizable, the matrices $\bar{R}$ is positive definite and $(\Gamma_A, \tilde{Q})$ is detectable. Then the ORL which stabilizes system (6.35) and minimizes $J(\infty)$ is given by*

$$\rho_k^* = -K(x(t_k) - \mathbf{x}_r) + \mathbf{z}_r, \quad k = 0, 1, \ldots,$$

where K is determined by

$$K = (\Gamma_B^T S \Gamma_B + \bar{R})^{-1}(\Gamma_B^T S \Gamma_A + \bar{T}^T) \tag{6.52}$$

and S is the solution of the Riccati equation

$$S = \tilde{Q} + \bar{\Gamma}_A^T S \bar{\Gamma}_A - \bar{\Gamma}_A^T S \Gamma_B (\Gamma_B^T S \Gamma_B + \bar{R})^{-1} \Gamma_B^T S \bar{\Gamma}_A. \tag{6.53}$$

Furthermore, the minimum of $J(N)$ is given by

$$J^*(\infty) = (x(0) - \mathbf{x}_r)^T S (x(0) - \mathbf{x}_r).$$

Remark 6.4. *For the finite horizon case, since (6.47) and (6.48) can only be solved backwards, it cannot be implemented online. However, for the infinite horizon case, if the reset time instants are equidistant, the reset law is of feedback form and thus it is very easy to be implemented online.*

6.4 Application to HDD systems

6.4.1 Dynamics model of HDD systems

The typical dynamics model of a hard disk drive *voice coil motor* (VCM) actuator is shown in Figure 6.1, where w_{in} and w_{out} are input disturbance and output disturbance, respectively. The dynamics of the nominal plant considered in this section is [2, 3]

$$\begin{cases} \dot{x}_1 = x_2, \\ \dot{x}_2 = -a x_1 - b x_2 + c u, \end{cases} \tag{6.54}$$

where

$$a = 1.4223 \times 10^5, \quad b = 3.016 \times 10^2, \quad c = 1.134 \times 10^7, \quad |u| \leq 3\text{V}.$$

The high-frequency resonance mode is given by

$$G(s) = G_1(s) G_2(s) G_3(s), \tag{6.55}$$

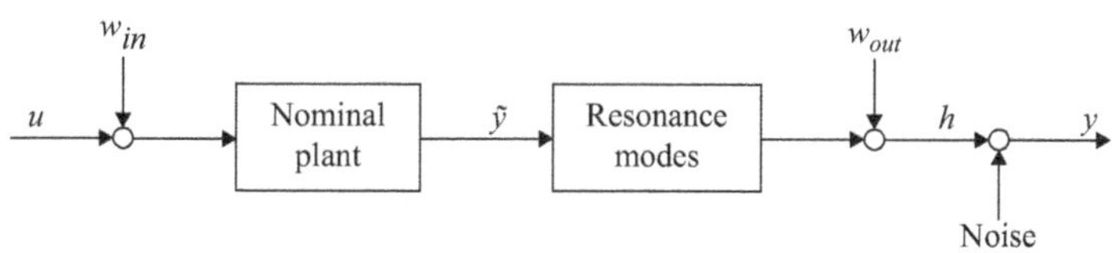

Figure 6.1 The model of VCM actuator

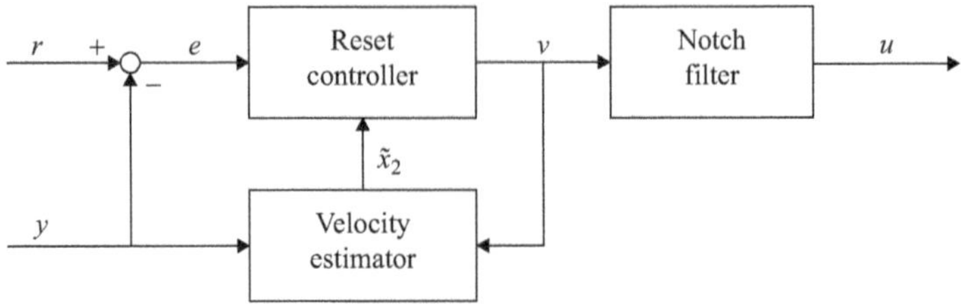

Figure 6.2 Structure of reset controller

where

$$G_1(s) = \frac{s^2 + 1.407 \times 10^4 s + 1.238 \times 10^9}{s^2 + 6.256 \times 10^4 s + 1.996 \times 10^9},$$

$$G_2(s) = \frac{s^2 + 1.508 \times 10^4 s + 1.421 \times 10^9}{s^2 + 3382s + 1.144 \times 10^9},$$

$$G_3(s) = \frac{s^2 + 2711s + 2.047 \times 10^9}{s^2 + 3821s + 1.469 \times 10^9}.$$

6.4.2 Moving horizon optimal reset control

6.4.2.1 Reset control design

The structure of the overall controller is as shown in Figure 6.2. The notch filter $G_N(s)$ is given by

$$G_N(s) = \frac{0.8594(s^2 + 4.812 \times 10^3 s + 8.588 \times 10^8)}{s^2 + 1.025 \times 10^4 + 8.911 \times 10^8} \times \frac{s^2 + 593.6s + 1.219 \times 10^9}{s^2 + 1.212 \times 10^4 + 1.012 \times 10^9} \tag{6.56}$$

which is designed to compensate for the high-frequency modes $G_1(s)$ and $G_3(s)$. The design process of a reset short-span-seeking controller is conducted on the nominal plant. For the base linear system, we first design the following controller

$$\begin{cases} v = \alpha e - \beta x_2 + \gamma z, \\ \dot{z} = e, \end{cases} \tag{6.57}$$

where

$$\alpha = 3, \quad \beta = 2 \times 10^{-3}, \quad \gamma = 3 \times 10^4.$$

Since the design is based on the nominal system, we do not take into account of high-frequency modes and notch filter in this stage. Thus we can identify v with u. In simulations, high-frequency modes and notch filter are inserted and u is replaced by

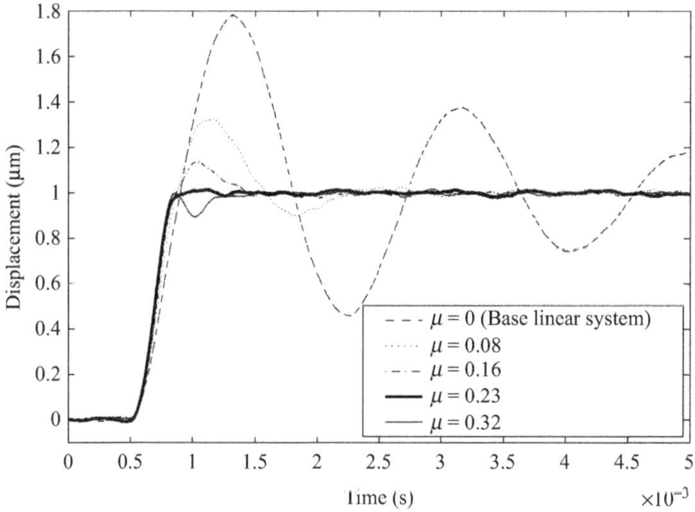

Figure 6.3 Step responses for different μ

$u = G_N(s)v$. The base linear system is designed to have rapid response and large bandwidth. Thus the transient response exhibits oscillations and overshoot (see Figure 6.3). We adopt the equidistant reset law, and the parameters of reset law are chosen as

$$P_0 = 1, \quad Q_0 = 6 \times 10^{-9}, \quad P_1 = 0.$$

Δt_k is chosen as 0.05 ms which is the same as the sampling period. Parameter μ is adjusted in simulations and we find that $\mu = 0.23$ provides very good performance. Thus, we obtain the following reset law

$$\begin{cases} \dot{z} = e, & t \neq t_k, \\ z(t^+) = a_1 x_1 - a_2 x_2 + a_3 z - a_4 r, & t = t_k, \end{cases} \tag{6.58}$$

where

$$a_1 = 7.0714 \times 10^{-5}, \quad a_2 = 8.8138 \times 10^{-9}, \quad a_3 = 0.77, \quad a_4 = 7.0647 \times 10^{-5}.$$

In practice, velocity x_2 is not measurable, thus we design an observer to estimate it. The estimator is given by

$$\begin{cases} \dot{\zeta} = -\lambda y - \tau \zeta + cv, \\ \tilde{x}_2 = \zeta + \nu y, \end{cases} \tag{6.59}$$

where

$$\lambda = 9.6842 \times 10^7, \quad \tau = 1 \times 10^4, \quad \nu = 9.6984 \times 10^3.$$

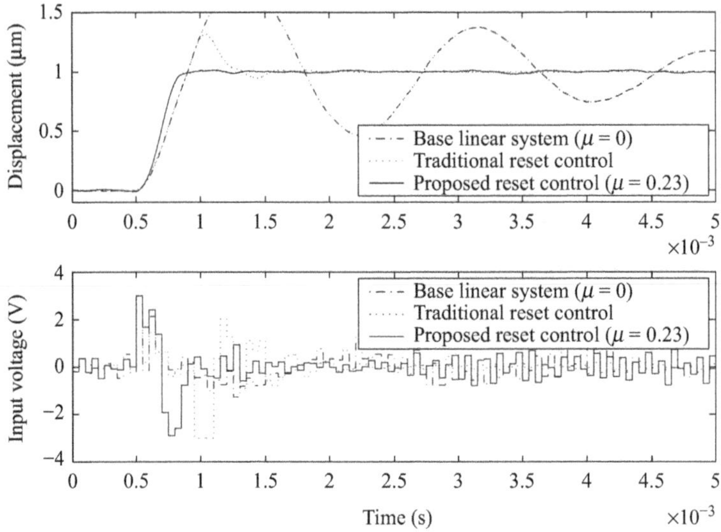

Figure 6.4 Step responses of base linear system, traditional reset control, and the proposed moving horizon optimal reset control

Replacing x_2 by $\tilde{x}_2$, the resulting controller is

$$\begin{cases} v = \alpha e - \beta\tilde{x}_2 + \gamma z, & \\ \dot{z} = e, & t \neq t_k, \\ z(t^+) = a_1x_1 - a_2\tilde{x}_2 + a_3z - a_4r, & t = t_k. \end{cases} \tag{6.60}$$

It is easy to show the closed-loop system is asymptotically stable.

6.4.2.2 Simulation results

The sampling frequency in our simulation is 20 kHz. And the input step occurs at 0.5 ms. In simulation, disturbance and noise are artificially added to simulate the real environment. They are generated by passing normally distributed random signals through disturbance models.

In Figure 6.3, the step responses for different μ are given. Other parameters are as stated in the previous subsection. We found that when $\mu \in [0, 0.65]$, the overall system is asymptotically stable. When μ varies from 0 to 0.23, both of the rise time and the overshoot decrease. In our simulation, $\mu = 0.23$ provides excellent short-span-seeking performance.

Figure 6.4 gives the step responses of the base linear system ($\mu = 0$), traditional reset control, and the proposed reset control with $\mu = 0.23$. One sees that the proposed reset control achieves a much better trade-off between rise time and overshoot than the traditional reset control. The 5% rise times of traditional reset control and the proposed reset control are, respectively, 0.39 ms and 0.32 ms, while the 5% settling times are, respectively, 0.97 ms and 0.32 ms. The settling time is shortened by 61.3%. In addition, the overshoot of the traditional reset control is 31%, but in the proposed reset control, the overshoot is almost zero. See Figure 6.5.

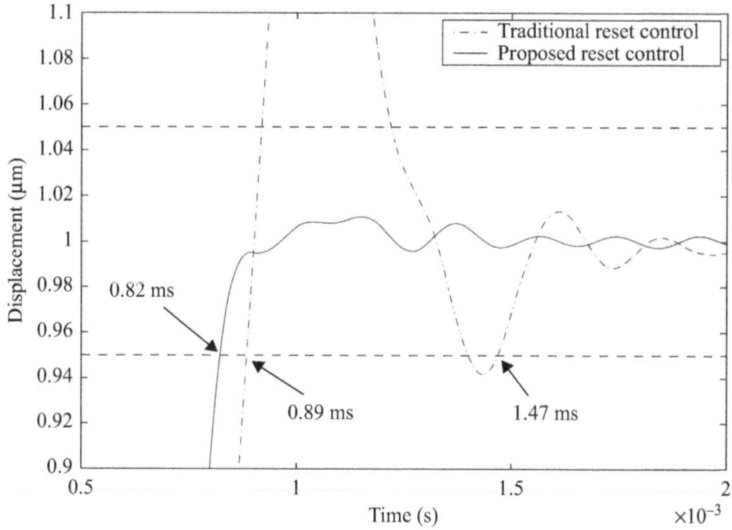

Figure 6.5 Enlarged view of step responses

6.4.3 Optimal reset control

In this section, we apply the optimal reset control to track-seeking control of a *hard disk drive* (HDD) system (6.54). Our aim is to design a seeking controller with fast positioning and without overshoot. The following is a guideline for the design:

1. First of all, we design a baseline controller such that the closed-loop system has a small damping ratio but without exceeding the input saturation. Since the reset is acting through the controller state, the baseline controller should be dynamic.
2. Second, choose an initial value of R/Q. Then, the ORL can be calculated directly by the method proposed in Section 6.3.2.
3. If the resulting response exhibits a large overshoot, then we increase the R/Q ratio and go to Step 2). If on the contrary, the resulting system is overdamped, we then try a smaller R/Q ratio and go to Step 2).
4. If adjusting R/Q ratio alone cannot achieve the desired performance, change the closed-loop damping ratio and go to Step 1). For instance, if overshoot cannot be removed by changing the R/Q ratio, reduce the damping ratio, and vice versa.

Generally, an acceptable controller can be obtained by following the procedure above.

6.4.3.1 Reset law design

The controller is depicted in Figure 6.2 where the reset controller is given by

$$\begin{cases} v = \alpha e - \beta x_2 + \gamma z & \\ \dot{z} = e, & t \neq t_k, \\ z(t^+) = \rho_k, & t = t_k, \end{cases} \tag{6.61}$$

where

$$\alpha = 1, \quad \beta = 2 \times 10^{-3}, \quad \gamma = 5 \times 10^{3}.$$

Thus, we have

$$\bar{A} = \begin{pmatrix} 0 & 1 & 0 \\ -(a + c\alpha) & -(b + c\beta) & c\gamma \\ -1 & 0 & 0 \end{pmatrix}.$$

If R/Q and the reset interval Δt are given, for instance,

$$Q = 10, \quad R = 0.3, \quad \Delta t = 5 \times 10^{-5}\text{s},$$

then the ORL can be derived. By (6.44), we have

$$\Gamma_A = \begin{pmatrix} 0.9890 & 2.9574 \times 10^{-5} \\ -3.8949 \times 10^{2} & 0.3093 \end{pmatrix},$$

$$\Gamma_B = \begin{pmatrix} 49.9128 \\ 1.6768 \times 10^{6} \end{pmatrix}.$$

Consider the infinite horizon case. By (6.41), (6.49), (6.50), and (6.51), we have $\bar{R} = 1.4808 \times 10^2$ and

$$\bar{\Gamma}_A = \begin{pmatrix} 0.9970 & 4.9533 \times 10^{-5} \\ -1.2099 \times 10^{2} & 0.9799 \end{pmatrix},$$

$$\tilde{Q} = \begin{pmatrix} 4.996 \times 10^{-4} & 1.2434 \times 10^{-8} \\ 1.2434 \times 10^{-8} & 4.1251 \times 10^{-13} \end{pmatrix}.$$

By solving Riccati equation (6.53) and (6.52), we obtain $K = (\, k_1 \quad k_2 \,)$ with

$$\begin{aligned} k_1 &= 1.1753 \times 10^{-3}, \\ k_2 &= 1.2898 \times 10^{-7}. \end{aligned} \tag{6.62}$$

Next, in view of Proposition 6.13, the ORL is given by

$$\rho_k^* = -k_1(x_1 - r) - k_2 x_2 + \frac{a}{c\gamma} r. \tag{6.63}$$

6.4.3.2 Simulation results

In simulation, the disturbance w_{in}, w_{out}, and noise are added to imitate the real environment. The high-frequency resonance models are given by (6.55). We use the same notch filter as in (6.56) and state estimator as in (6.59). In reset controller (6.61), x_2 is replaced by the estimated value $\tilde{x}_2$ in simulation.

Short-span seeking
The sampling period is $T_s = 5 \times 10^{-5}$ s in simulation and $R/Q = 0.03$. Figure 6.6 gives the step responses for the base linear system and the optimal reset control with reset interval $\Delta t = T_s$ (the step input occurs at $t = 5 \times 10^{-4}$ s). In this case, the base linear controller is given by the first two equations of (6.61) and the reset law is given by (6.63) and (6.62). We can see that the proposed optimal reset control almost removes the overshoot completely and at the same time reduces the rise time significantly.

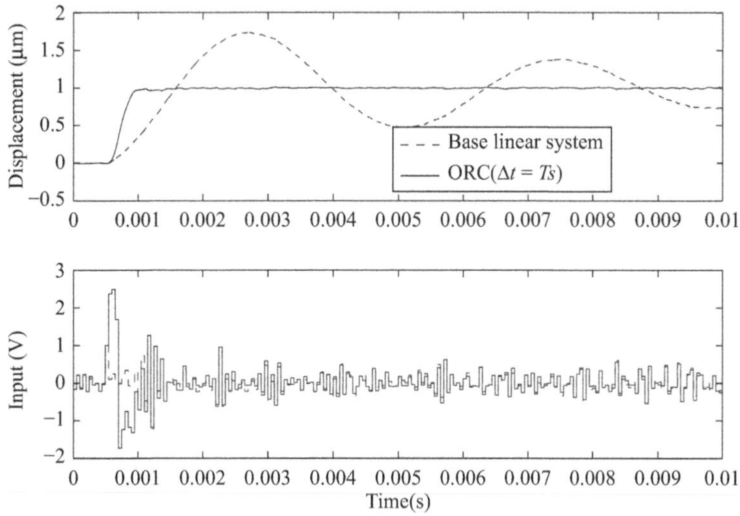

Figure 6.6 Step responses for base linear system and optimal reset control

Figure 6.7 gives the responses for different reset intervals $\Delta t = T_s$, $5T_s$, and $10T_s$, respectively. Obviously, the transient performance is better when reset interval is smaller. In Figure 6.8, step responses for different values of R/Q are given. Note that a smaller R/Q ratio leads to faster response, but with a higher input level, thus the control input is more likely to saturate. In this simulation, we can see that when $R/Q = 0.025$, the control input does not saturate and the transient performance is perfect.

Compared with the moving horizon optimization method in Section 6.2, the main advantage of the optimal reset control is that the design process is much simpler. We do not need to worry about the stability since the stability is naturally assured.

Long-span seeking

In long-span-seeking control, we choose $\Delta t = 3T_s$ and $R/Q = 0.3$. In this case, it is easy to calculate that the reset law is given by (6.63) with

$$\begin{aligned} k_1 &= 3.7092 \times 10^{-4}, \\ k_2 &= 1.6476 \times 10^{-7}. \end{aligned} \tag{6.64}$$

The input step signals all occur at 5×10^{-4} s. Figure 6.9 gives the responses for different step levels. We found that in the range $r \geq 25\mu$m, the fixed value $R/Q = 0.3$ achieves perfect long-span-seeking control.

From the simulation, one sees that the control input is almost Bang-bang during the rising stage (see the control input signals in Figures 6.10 and 6.11). Thus the long-span-seeking control is almost time-optimal.

Remark 6.5. *In the simulation, we found that the performance is not sensitive to* Δt *as long as* $\Delta t \in [T_s, 5T_s]$. *But if* $\Delta t > 5T_s$, *the performance would be worsened.*

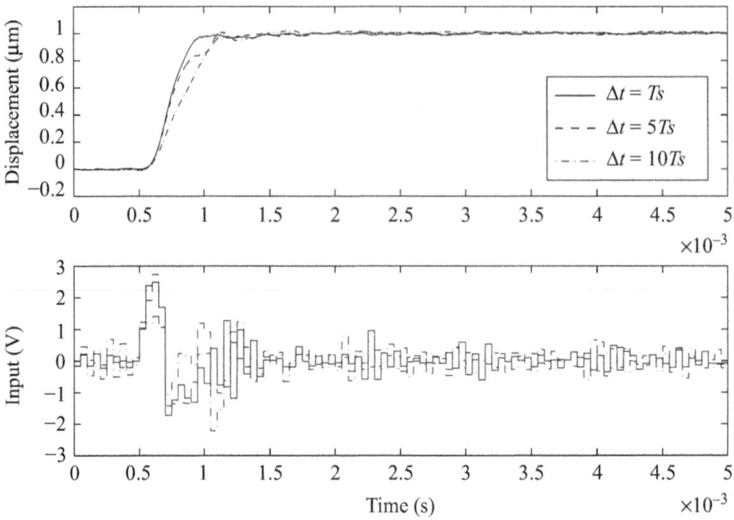

Figure 6.7 Step responses for optimal reset control with different Δt

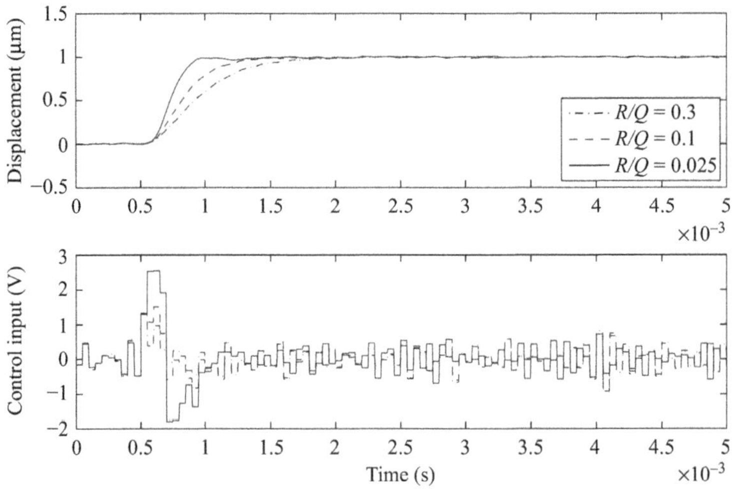

Figure 6.8 Step responses for different R/Q

In short-span seeking, we need different values of R/Q *for different reference levels* r *in order to achieve the best performance. However, in long-span-seeking control, the performance is not sensitive to* R/Q.

6.4.3.3 Comparison with composite nonlinear feedback control

For comparison, we design a *composite nonlinear feedback* (CNF)- seeking controller [4, 5] for the step level 200μm. Follow the idea of CNF design, we first design a

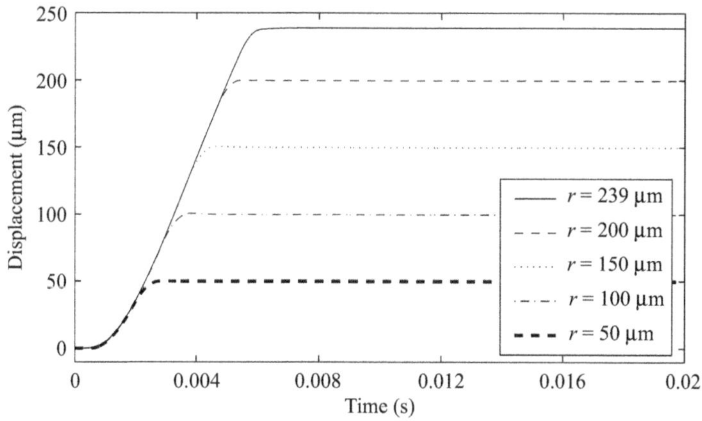

Figure 6.9 Responses for different step levels

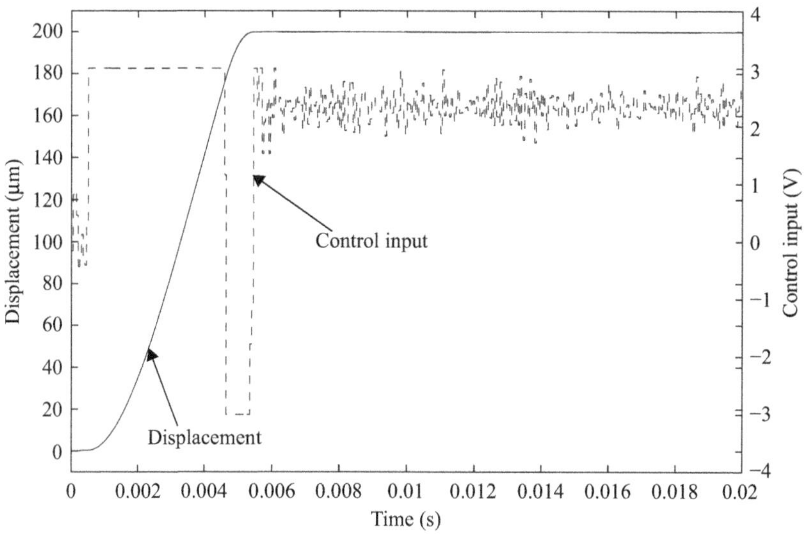

Figure 6.10 Response and control input for $r = 200\ \mu m$

linear feedback controller with a small damping ratio to achieve fast response during acceleration. The linear controller is given by

$$u_L = F\tilde{x} + Gr$$

where

$$F = \begin{pmatrix} -3.9742 \times 10^{-2} & 2.2522 \times 10^{-5} \end{pmatrix},$$
$$G = -(C(A+BF)^{-1}B)^{-1} = 5.2284 \times 10^{-2}.$$

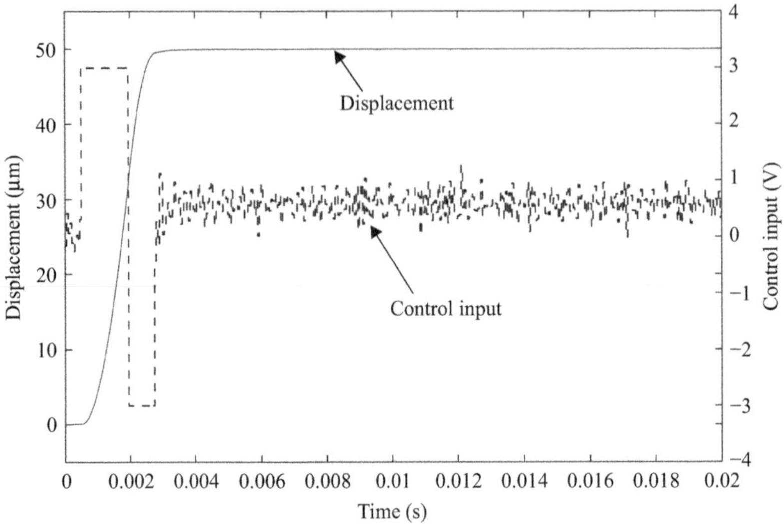

Figure 6.11 Response and control input for $r = 50\,\mu m$

and $\tilde{x} = (x_1, \tilde{x}_2)^T$ with $\tilde{x}_2$ the estimated value of x_2. The estimator is given by

$$\begin{cases} \dot{\xi} = (-b+k)\xi + (-a+bk-k^2)y + cu, \\ \tilde{x}_2 = \xi - (-b+k)y \end{cases}$$

with $k = -5000$. Then the nonlinear part of the form

$$u_N = \rho(|r-y|)F_N \begin{pmatrix} y-r \\ \tilde{x}_2 \end{pmatrix}$$

with

$$\rho(|r-y|) = -\beta_0 \left(e^{-\alpha_0|r-y|} + \gamma_0\right)$$

is designed such that the dominant poles of

$$A + BF - \beta_0(1+\gamma_0)BF_N$$

have a large damping ratio. Here, F_N is chosen as

$$F_N = \begin{pmatrix} 4.7862 \times 10^{-3} & 7.9537 \times 10^{-5} \end{pmatrix}$$

and the parameters β_0, α_0, and γ_0 are adjusted to achieve fast positioning without overshoot. For step level $r = 200\mu$m, we found that $\beta_0 = 4$, $\alpha_0 = 0.07$, and $\gamma_0 = 0.05$ yield good seeking performance.

The optimal reset control used for comparison is the same as the one in Section 6.4.3. The output responses and the control inputs of both the optimal reset control and *composite nonlinear feedback control* (CNFC) are given in Figures 6.12 and 6.13. From the simulation results, we can see that both the optimal reset control and CNF design can achieve perfect seeking performance by suitable choice of parameters.

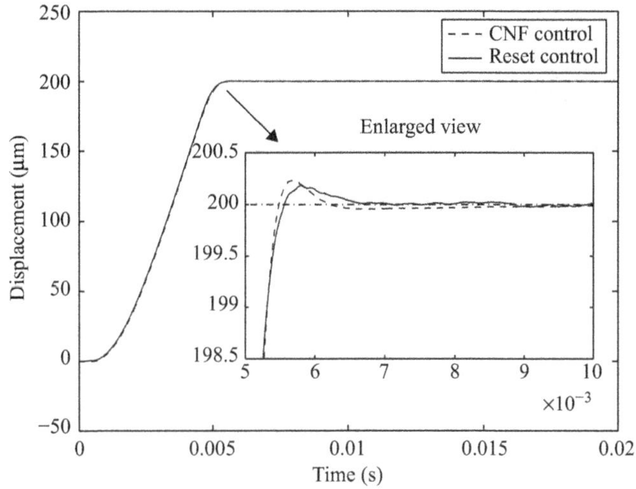

Figure 6.12 Output responses for both optimal reset control and CNFC for $r = 200\ \mu m$

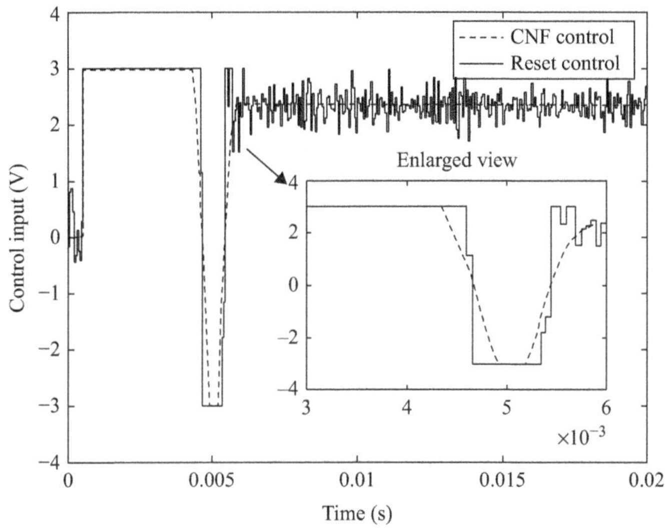

Figure 6.13 Control inputs for both optimal reset control and CNFC

We point out that, though from Figure 6.13, the reset control is closer to the Bang-bang control in this case, it is hard to say that which technique outperforms the other since both have rooms for improvement by optimizing the parameters. The advantage of the optimal reset control is that the design procedure is relatively simple. For instance, we do not need to design a transitional nonlinear function to smoothly connect two

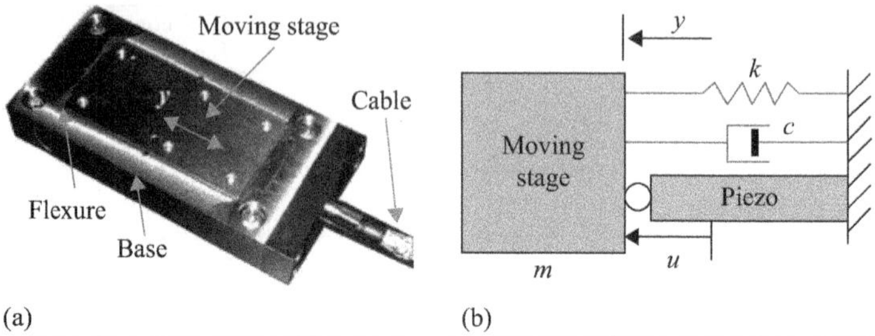

Figure 6.14 Piezoelectric (PZT) mircoactuator-positioning stage. (a) PZT-positioning state and (b) the mass-damper-spring model

different controllers which are designed separately. Furthermore, we found that the number of parameters to be adjusted are relatively small in the proposed reset control design.

6.5 Application to PZT-positioning stage

6.5.1 Modeling of the PZT-positioning stage

The PZT-positioning stage consists of a PZT microactuator, a moving stage connected with the base via the flexures in four corners, a PZT power amplifier and an integrated capacitive position feedback sensor with 0.2 nm resolution to measure the displacement of the moving stage. The PZT microactuator is of high stiffness and has a maximum travel range of $\pm 15\,\mu$m. The mechanical resonance caused by the flexures is actively damped by the integrated control electronics. Thus, the dynamics of the PZT-positioning stage can be simply depicted by a mass-damper-spring system as shown in Figure 6.14(b), which can be described by a state-space form as follows

$$\Sigma_p : \quad \begin{cases} \dot{x}_1 = x_2 \\ \dot{x}_2 = -a_1 x_1 - a_2 x_2 + bu \\ y = x_1 \end{cases} \tag{6.65}$$

where x_1 and x_2 are the position and velocity of the moving stage, respectively, and u is the control input to the PZT amplifier.

The modal parameters in (6.65) are identified from experimental frequency response data. A *dynamic signal analyzer* (DSA) (HP 35670A, Hewlett Packard Company, Washington) is used to generate the swept-sinusoidal excitation signals and collect the frequency response data from the excitation signals u to the position output y. The dashed lines in Figure 6.15 show the measured frequency responses of the PZT-positioning stage. The PZT dynamics is of high stiffness that exhibits a flat gain in the low-frequency range. Using the least square estimation method [6], we obtain the modal parameters as follows

$$a_1 = 10^6, \quad a_2 = 1810, \quad b = 3 \times 10^6.$$

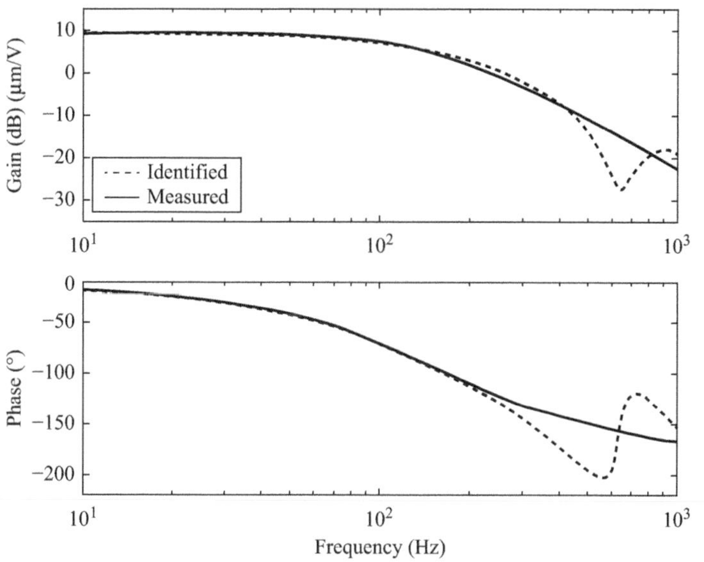

Figure 6.15 Frequency responses of the PZT microactuator-positioning stage

The solid lines in Figure 6.15 show that the identified model matches the measured one well in the frequency range of interest.

6.5.2 Reset control design

Our objective is to design a feedback controller for robust tracking of a step reference input with zero steady-state error and fast settling, for which high open-loop gains in low-frequency range and high bandwidth with sufficient stability margin are typically required. Thus, we employ the classical *proportional-plus-integral* (PI) control structure as the base controller

$$\frac{u(s)}{e(s)} = k_p + \frac{k_i}{s} \tag{6.66}$$

where $k_p = 0.08$ and $k_i = 300$. The resultant base linear system has a closed-loop bandwidth 178 Hz, and gain/phase margin 11 dB/33°. Experimental result (see the dashed lines in Figure 6.16) shows that the integrator increases the low-frequency gains and both fast rise time and zero steady-state error are achieved. However, the overshoot induced by the integrator is also significant (40%), which results in tedious settling time. Therefore it is expected that the overshoot could be reduced by resetting the integrator state to proper values.

To design the reset values, we set the reset time interval as a constant $\Delta t_k = 1$ ms and select the tuning parameters of J_k in (6.12) as

$$P_0 = 2.1, \quad Q_0 = 1 \times 10^{-6}, \quad P_1 = 0. \tag{6.67}$$

Thus, according to (6.13) and (6.20) we can easily obtain $\mathbf{x}_r$ and Γ as

$$\mathbf{x}_r = \begin{bmatrix} 1 \\ 0 \\ 0.0011 \end{bmatrix}, \quad \Gamma = \begin{bmatrix} 1.16 & 0.0005 & 65.33 \\ 0.0005 & 0 & 0.16 \\ 65.33 & 0.16 & 230042 \end{bmatrix}.$$

The moving horizon *optimal reset controller* (moving horizon ORC) is then described in state space as follows

$$\begin{cases} \dot{x}_r = e, & t \neq t_k, \\ x_r^+ = E_1 x_1 + E_2 x_2 + Gr, & t = t_k, \\ u = k_i x_r + k_p e \end{cases} \tag{6.68}$$

where

$$E_1 = -2.8 \times 10^{-4}, \quad E_2 = -6.8 \times 10^7,$$

$$G = 0.0014, \quad k_i = 300, \quad k_p = 0.08.$$

Moreover, it is easy to verify that the resulting RCS satisfies the stability condition in (6.24), which implies that the unforced closed-loop system is asymptotically stable. For comparison, we also design a conventional reset controller as follows

$$\begin{cases} \dot{x}_r = e, & e(t) \neq 0, \\ x_r^+ = 0, & e(t) = 0, \\ u = 300x_r + 0.8e. \end{cases} \tag{6.69}$$

In fact, the above conventional reset controller can be seen as a special case of the proposed reset controller. Thus, the stability of the closed-loop system under this controller can be easily verified through (6.24).

6.5.3 Experimental results

The reset controllers were implemented by a real-time *digital signal processor* (DSP) system (dSPACE-DS1103) with the sampling time of $T_s = 50$ μs. The position of the moving stage x_1 or equivalently the output y can be directly obtained through the sensor output. We estimate the velocity x_2 by backward differentiation of the position signal $x_1(t)$, i.e.,

$$\hat{x}_2 = \frac{z-1}{T_s z} x_1.$$

Figure 6.16 shows the experimental results for 1 μm step response. It can be seen that the extended reset control nearly removes the overshoot and thus reduces the settling time from 15 ms (no reset) to 3 ms. Moreover, we observe that the extended reset control has a faster transient response compared with that under no reset because the integrator state was reset to minimize J_k from the beginning at $t = 0$ leading to a larger control input at the initial stage and thus faster response. When the position output approaches the target, the integrator state is reset to a smaller value (see the control input at $t = 1$ ms in Figure 6.16 to reduce the overshoot and keep the moving stage at the desired position. In this case, the conventional reset control works badly,

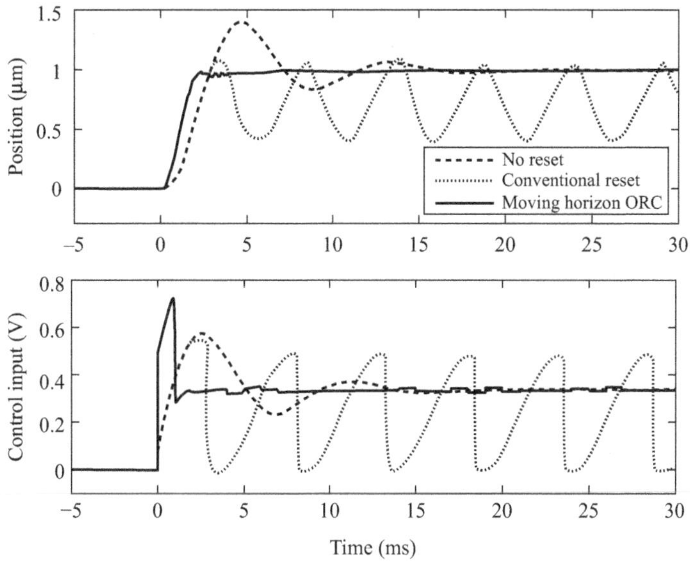

Figure 6.16 Step responses ($r = 1\ \mu m$)

which can only partially reduce the first overshoot peak and results in limit cycles. This is because resetting the integrator state to zero tends to resetting the control input to zero, which will cause the moving stage going to its initial position due to the high stiffness of the PZT actuator. Thus, the conventional reset control needs an intentional interplay between the reset mechanism and an appropriately designed base linear controller [7].

Next, we test the step responses of the extended RCS to various step levels. The results are shown in Figure 6.17, which indicates that the overshoots in all cases are nearly removed and the settling time is still maintained to be 3 ms. In this chapter, we have not theoretically considered how to select the reset time interval, which is in fact related to the overall system performance. This is however evaluated through experiments. To do this, we use the same design parameters (6.67) but select various reset time intervals. Following the same design procedure, we obtain a set of reset controller for implementation. The results in terms of net overshoot and settling time are summarized and compared in Figure 6.18. It is interesting to see that a smaller reset time interval tends to exhibit less overshoot but results in longer settling time (i.e., slower system dynamics). Under the range of 0.4–2 ms, the settling time and overshoot simultaneously reach a low level. Thus, tuning the reset time interval as 20–60% of the rise time (3 ms in this case) by the base linear system achieves relatively good performance.

Finally, we test the robustness of the moving horizon optimal RCS against input disturbance and sensor noise. Figures 6.19 and 6.20, respectively, show the time responses to a single-frequency (100 Hz) sinusoidal input disturbance and sensor noise, which are artificially introduced to the control system. We can see that the

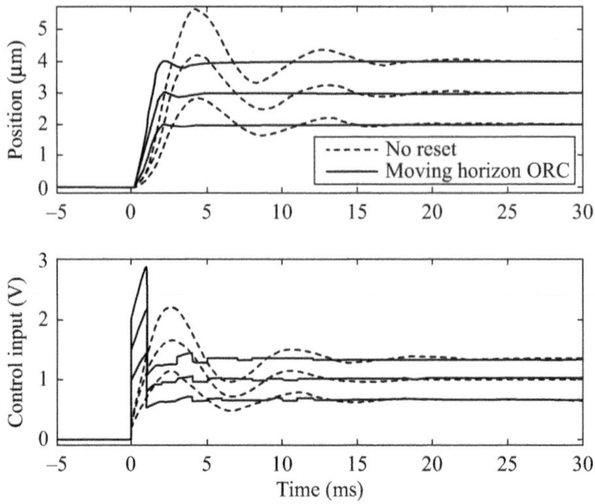

Figure 6.17 Time responses for various step levels ($r = 2, 3, 4\ \mu m$)

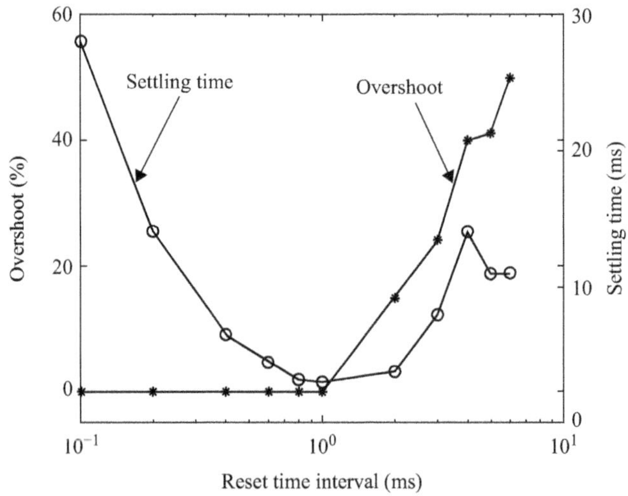

Figure 6.18 Relationship between reset time interval, overshoot, and settling time of the moving horizon optimal RCSs

moving horizon optimal reset control simultaneously provides an improvement of 65% in both disturbance and noise suppression based on the base linear system (no reset). Further, we experimentally analyze the properties of disturbance and noise suppression in a wide frequency range according to the *describing function* (DF) approach. We used a DSA to generate swept-sinusoidal excitation signals, which are then injected to the control input or the sensor output, respectively. The tests of disturbance and sensor noise responses are performed individually. The DSA is also

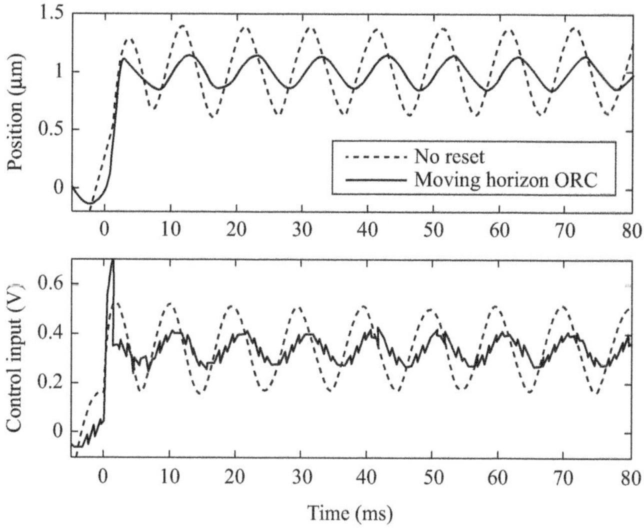

Figure 6.19 Time responses to step input $r = 1\ \mu m$ and sinusoidal input disturbance $u_d = 0.1\ sin\ (200\pi t) V$

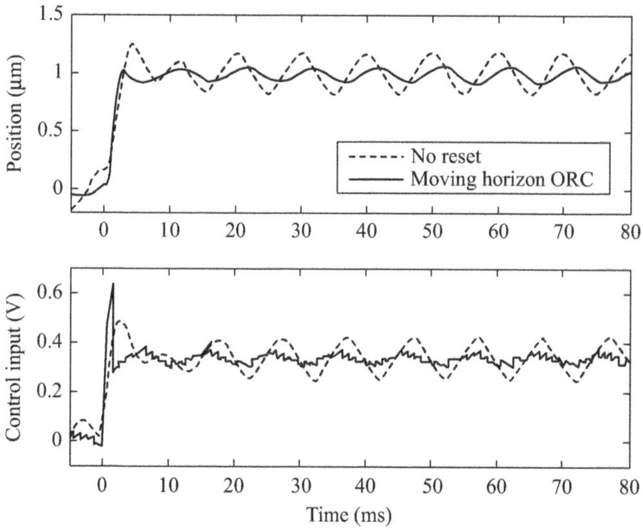

Figure 6.20 Time responses to step input $r = 1\ \mu m$ and sinusoidal sensor noise $n = 0.1\ sin\ (200\pi t) \mu m$

used to collect the frequency response data from the position output to the excitation signals. Since the RCS is essentially nonlinear, its frequency response may depend on the excitation levels. We thus vary the excitation level from 0.1 to 3 evenly spaced by 0.1, and it is interesting to find that the Bode plots of the frequency responses in

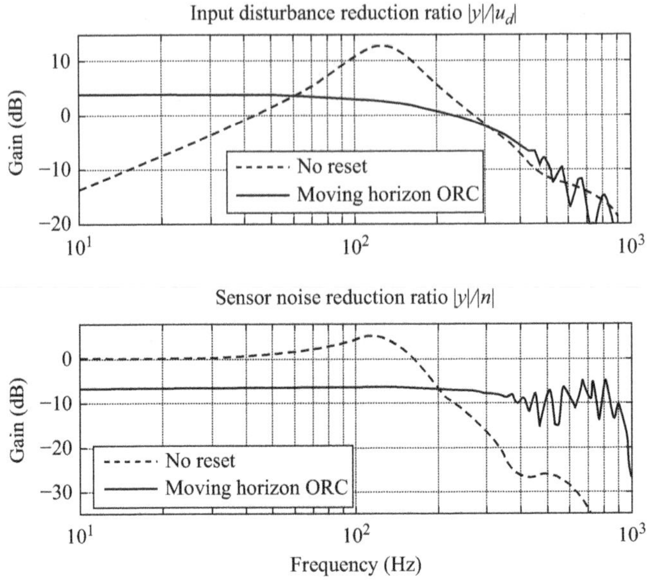

Figure 6.21 Bode plot for DFs of input disturbance/sensor noise suppression; the excitation level of u_d is 0.1 V and n is 0.1 μm

all cases are very close to that in Figure 6.21. In the plot, we can see that improved disturbance/noise suppression occurs simultaneously around 100 Hz, which matches the results in Figures 6.19 and 6.20. However, we also find that the moving horizon optimal reset control adversely increases the low-frequency disturbance (< 60 Hz) and high-frequency noise (> 200 Hz) reduction ratio.

Notes

The moving horizon optimal reset control and the infinite horizon optimal reset control design of this chapter are, respectively, based on References 3 and 8. The application of the moving horizon ORC to PZT-positioning stage is based on Reference 9. Propositions 6.1 and 6.2 are based on the common Lyapunov function condition and the Lie-algebraic criteria for stability of switched system in References 1, 10, 11, and 12. The solvability of the LQR problem for discrete-time systems can be found in Reference 13, for instance. The ORL designs proposed in this chapter are closely related to the optimal impulsive system design in Reference 14 where a nonlinear model was considered and a numerical algorithm for optimal impulse time instants and values has been given aiming to save computation time.

There are many other techniques which have been developed and applied to improve track-seeking performance of HDD systems. They include time optimal control [15], [16], sliding mode control [17], and CNFC [4], [18], etc.

References

[1] D. Liberzon, J.P. Hespanha, and A.S. Morse. Stability of switched systems: a lie-algebraic condition. *Systems & Control Letters*, 37(3):117–122, 1999.

[2] Y. Li, G. Guo, and Y.Wang, Phase lead reset control design with an application to HDD servo systems, *The 9th International Conference on Control, Automation, Robotics and Vision*, 5–8 December 2006, IEEE, Singapore, pp. 1–6.

[3] Y. Guo, Y. Wang, L. Xie, and J. Zheng. Stability analysis and design of reset systems: theory and an application. *Automatica*, 45(2):492–497, 2009.

[4] B.M. Chen, T.H. Lee, K. Peng, and V. Venkataramanan. Composite nonlinear feedback control for linear systems with input saturation: theory and an application. *IEEE Transactions on Automatic Control*, 48(3):427–439, 2003.

[5] V. Venkataramanan, K. Peng, B.M. Chen, and T.H. Lee. Discrete-time composite nonlinear feedback control with an application in design of a hard disk drive servo system. *IEEE Transactions on Control Systems Technology*, 11(1):16–23, 2003.

[6] L. Ljung. System identification: theory for the user. *In Information and System Sciences Series*, vol. 198, Prentice Hall PTR, Upper Saddle River, NJ, 1987.

[7] Y. Zheng, Y. Chait, C.V. Hollot, M. Steinbuch, and M. Norg. Experimental demonstration of reset control design. *Control Engineering Practice*, 8(2):113–120, 2000.

[8] Y. Guo, Y. Wang, L. Xie, H. Li, and W. Gui. Optimal reset law design and its application to transient response improvement of HDD systems. *IEEE Transactions on Control Systems Technology*, 19(5):1160–1167, 2011.

[9] J. Zheng, Y. Guo, M. Fu, Y. Wang, L. Xie, Development of an extended reset controller and its experimental demonstration. *IET Control Theory Applications*, 2(10), 866–874, 2008.

[10] K.S. Narenda and J. Balakrishnan. A common Lyapunov function for stable LTI systems with commuting A-matrices. *IEEE Transactions on Automatic Control*, 39(12):2469–2471, 1994.

[11] D. Cheng, L. Guo, and J. Huang. On quadratic lyapunov functions. *IEEE Transactions on Automatic Control*, 48(5):885–890, 2003.

[12] A.A. Agrachev and D. Liberzon. Lie-algebraic stability criteria for switched systems. *SIAM Journal on Control and Optimization*, 40(1):253–269, 2001.

[13] V. Strejc. State space theory of discrete linear control. John Wiley & Sons, Inc. New York, NY, USA, 1981.

[14] K.H. Kyung. Impulse control: Boolean programming and numerical algorithms. *IEEE Transactions on Automatic Control*, 51(8):1230–1248, 2006.

[15] B.K. Kim, W.K. Chung, H.S. Lee, H. Choi, I.H. Suh, and Y.H. Chang. Robust time optimal controller design for hard disk drives. *IEEE Transactions on Magnetics*, 35(5):3598–3600, 1999.

[16] M.L. Workman, R.L. Kosut, and G.F. Franklin. Adaptive proximate time optimal servomechanisms-continuous time case. *The 6th American Control Conference*, June 10–12, 1987, IEEE, Minneapolis, MN, pp. 589–594, 1987.

[17] D.Q. Zhang and G.X. Guo. Discrete-time sliding mode proximate time optimal seek control of hard disk drives. *IEE Proceedings-Control Theory and Applications*, 147(4):440–446, 2000.

[18] B.M. Chen and W. Lan. On improving transient performance in tracking control for a class of nonlinear discrete-time systems with input saturation. *IEEE Transactions on Automatic Control*, 52(7):1307–1313, 2007.

Chapter 7
Reset control systems with conic jump sets

This chapter discusses *reset control systems* (RCSs) with conic jump sets. Different from the zero-crossing type reset where a jump set is a subset with zero measure, a conic jump set is a closed conic in the state space or the input/output space which contains a nonempty interior. The key role a conic jump set plays is to remove part of the state space and force the trajectories of a system to evolve within the remaining part. This kind of reset leads to less conservative Lyapunov stability conditions and is capable of making a system passive. This chapter only briefly introduces the main idea and collects some of the main results reported in recent literature including passification via reset and $\mathcal{L}_2$-gain analysis.

7.1 Basic idea

In most of the early literature, the *Clegg integrator* (CI) is modeled by the zero-crossing type reset element

$$\begin{cases} \dot{x}_r = v, & v \neq 0, \\ x_r^+ = 0, & v = 0, \end{cases} \tag{7.1}$$

i.e., whenever the input of the integrator crosses zero, the accumulated state is reset to zero. The benefits that this zero-crossing type CI brings have been explained in Chapters 1 and 2 through its sinusoid input *describing function* (DF). However, the zero-crossing type reset also has some restrictions. For example, the jump set for a zero-crossing type RCS is a subset with measure zero. This feature leads to the condition that the Lyapunov function constructed to establish the stability results has to be well-defined on almost the whole state space and thus the stability of the baseline system is required in general. As pointed out in Chapter 2, a stable zero-crossing RCS might contain an unstable baseline system. For this kind of systems, Lyapunov-function-based stability criterion is difficult to obtain. As far as the stability analysis is concerned, the reset times-dependent stability results presented in Chapter 2 apply to this situation. However, it is not convenient for the control design of reset systems. Another restriction of the zero-crossing type reset is that it is difficult to be extended to multi-input multi-output case.

A different reset scheme which is more natural to model the original proposal of Clegg is [1]

$$\begin{cases} \dot{x}_r = v, & x_r v \geq 0, \\ x_r^+ = 0, & x_r v \leq 0. \end{cases} \tag{7.2}$$

In this case, the state of the integrator is reset to zero whenever the input and the output are of opposite signs. Similarly, a *first-order reset element* (FORE) with this kind of reset is given by

$$\begin{cases} \dot{x}_r = \lambda x_r + v, & x_r v \geq 0, \\ x_r^+ = 0, & x_r v \leq 0. \end{cases} \tag{7.3}$$

Interestingly, when the initial state is zero, i.e., $x_r(0^+) = 0$, the zero-crossing type FORE and the FORE with conic jump set (7.3) have the same output response to a continuous excitation input. Actually, the output response of the baseline system

$$\dot{x}_r = \lambda x_r + v$$

is given by

$$x_r(t) = \int_0^t e^{\lambda(t-s)} v(s) ds$$

which is of the same sign with $v(t)$ before the first zero-crossing of $v(t)$, i.e.,

$$x_r(t)v(t) \geq 0, \quad \forall t \in (0, \tau_1],$$

where τ_1 is the smallest positive time instant such that $v(t)$ changes its sign. Thus every reset action occurs exactly at the zero-crossing of $v(t)$. Based on this observation, the zero-crossing type FORE and the FORE with conic jump set (7.3) share the common sinusoid input DF. However, when the initial state is nonzero, or the excitation input is discontinuous, the output responses of these two FOREs are different. In the following, we discuss what different features this conic-type reset brings. Consider an RCS consisting of the FORE (7.3) and a linear plant

$$\begin{cases} \dot{x}_p = A_p x_p + b_p u \\ y_p = c_p x_p \end{cases} \tag{7.4}$$

with

$$u = x_r, \quad v = -y.$$

The closed-loop system of this RCS is given by

$$\begin{cases} \dot{x} = Ax, & x \in \mathcal{F}, \\ x^+ = Rx, & x \in \mathcal{J}, \end{cases} \tag{7.5}$$

where $x = (x_p^T, x_r)^T$,

$$\mathcal{F} = \{x \mid x^T M x \geq 0\}, \tag{7.6}$$

$$\mathcal{J} = \{x \mid x^T M x \leq 0\}, \tag{7.7}$$

$$A = \begin{bmatrix} A_p & b_p \\ -c_p & \lambda \end{bmatrix}, \quad R = \begin{bmatrix} I_{n_p} & 0 \\ 0 & 0 \end{bmatrix}, \quad M = \begin{bmatrix} 0 & -c_p^T \\ -c_p & 0 \end{bmatrix}.$$

One sees that in the above RCS, the flow set $\mathcal{F}$ and the jump set $\mathcal{J}$ are both closed subsets (cones) and are allowed to overlap, i.e.,

$$\mathcal{F} \cap \mathcal{J} \neq \emptyset.$$

These features allow us to put this kind of RCSs into the framework of hybrid systems proposed in Reference 2. Actually, more general reset conditions can be defined as

$$\mathcal{F} = \{x \mid x^T F x \geq 0\}, \tag{7.8}$$

$$\mathcal{J} = \{x \mid x^T J x \geq 0, \ x \neq 0\}, \tag{7.9}$$

where F and J are symmetric matrices and $\mathcal{F} \cap \mathcal{J}$ contains nonempty interior. The overlap of the flow set and the jump set provides convenience to characterize robustness with respect to uncertain reset time instants. For instance, when time-regularization is applied, the flow set would slightly enter the interior of the jump set.

An important role that the conic jump set plays is to remove part of the state space and force the trajectory of the system to evolve within a subset of the whole state space. This helps to establish less conservative Lyapunov-function-based stability criteria because the Lyapunov function is not required to be well-defined on the whole state space. For instance, consider an RCS which is of the form (7.5) with $\mathcal{F}$ and $\mathcal{J}$ as defined in (7.8) and (7.9), respectively, where M is a symmetric matrix. Suppose that

$$R(\mathcal{J}) \subseteq \mathcal{F}^{\circ} \tag{7.10}$$

where $\mathcal{F}^{\circ}$ represents the set of interiors of $\mathcal{F}$. Then, the Zeno phenomenon will not occur. In this case, stability is assured if there is a Lyapunov function candidate $V(x)$ which is only needed to be well-defined and positive definite within the flow set $\mathcal{F}$, i.e.,

$$V(x) > 0, \quad x \in \mathcal{F}, \tag{7.11}$$

such that it decreases along any trajectory starting from the flow set, i.e.,

$$\dot{V}(x) < 0, \quad \forall x \in \mathcal{F}, \tag{7.12}$$

$$V(Rx) - V(x) \leq 0, \quad \forall x \in \mathcal{F} \cap \mathcal{J}. \tag{7.13}$$

Note that the inequality (7.13) is only required to hold within $\mathcal{F} \cap \mathcal{J}$ instead of the whole jump set $\mathcal{J}$. This is because the only situation that any trajectory intersects the subset $\mathcal{J} \setminus \mathcal{F}$ is $x(0) \in \mathcal{J} \setminus \mathcal{F}$. In this case, one has

$$x(0^+) = Rx(0) \in \mathcal{F}^{\circ}.$$

Since the reset mapping defined by the reset matrix R is linear, we can use $x(0^+)$ as the initial state in any case without losing any generality. In other words, the initial state can be restricted to be within the flow set in which case any trajectory will never go into $\mathcal{J} \setminus \mathcal{F}$. Thus the subset $\mathcal{J} \setminus \mathcal{F}$ can be removed from consideration. For instance, if

$F = J = M$ and we restrict our attention to a quadratic Lyapunov function candidate of the form

$$V(x) = x^T P x, \quad x \in \mathcal{F},$$

then by using $\mathcal{S}$-procedure [3], the set of inequalities (7.11), (7.12), and (7.13) hold if there are positive numbers μ_1, μ_2, and symmetric matrix P such that

$$P - \mu_1 M > 0, \tag{7.14}$$

$$A^T P + PA + \mu_2 M < 0, \tag{7.15}$$

$$x^T (R^T P R - P) x \leq 0, \ \forall x \text{ s.t. } x^T M x = 0. \tag{7.16}$$

One of the natural extensions of (7.3) to MIMO systems is

$$\begin{cases} \dot{x}_r = A_r x_r + B_r v, & v^T y \geq 0, \\ x_r^+ = R_r x_r, & v^T y \leq 0, \\ y = C_r x_r + D_r v, \end{cases} \tag{7.17}$$

i.e., the reset action is triggered whenever the input vector $v(t)$ and output vector $y(t)$ form an obtuse angle. Intuitively, this kind of reset forces the trajectories of the system to evolve within the first and the third quadrants $\{(v, y) \mid v^T y \geq 0\}$ of the input/output space (for single-input single-output case), or in other words, the part $\{(v, y) \mid v^T y < 0\}$ has been removed from the input/output space. Integrate $v^T y \geq 0$ gives

$$\int_0^T v^T(s) y(s) ds \geq \beta(x_r(0)) := v^T(0) y(0)$$

which implies that the reset makes the system (7.17) passive. This point will be revisited in the next section.

7.2 L_2-gain analysis

First of all, we introduce some notations regarding *hybrid systems* used in this section, which are taken from References 4 and 5. A subset of $[0, \infty) \times \mathbb{Z}_{\geq 0}$ is called a *hybrid time domain* if it can be expressed as

$$\bigcup_{i=1}^{N} [t_i, t_{i+1}] \times \{i\}$$

where $N \in \mathbb{Z}_{\geq 0} \cup \{\infty\}$, $0 = t_0 \leq t_1 \leq \cdots$ and the last interval is allowed to be of the form $[t_N, T)$ with T finite or $T = +\infty$. A n-dimensional hybrid signal ξ is a function defined on a hybrid time domain, denoted by dom(ξ), i.e., $\xi : \text{dom}(\xi) \to \mathbb{R}^n$. A *hybrid signal* ξ is called a *hybrid arc* if $\xi(t, j)$, $t \in [t_j, t_{j+1}]$, is locally absolutely continuous with respect to t for any j. For a given hybrid signal η defined on hybrid time domain dom (η), define a continuous-time signal $\bar{\eta}(t)$ as

$$\bar{\eta}(t) = \begin{cases} \eta(t, j), & (t, j) \in \text{dom}(\eta), (t, j+1) \notin \text{dom}(\eta), \\ \eta(t, j+1), & \text{otherwise.} \end{cases}$$

$\bar{\eta}$ is called the projection from η on $\mathbb{R}_{\geq 0}$. For any continuous-time signal $\alpha(t)$, any hybrid signal η satisfying $\bar{\eta} = \alpha$ is called a hybrid signal lifted from $\alpha(t)$. For any $(t,j) \in \text{dom}(\eta)$ and any $p \in [1, \infty)$, define the integral

$$\int_0^t \eta(s)ds := \sum_{i=0}^{j-1} \int_{t_i}^{t_{i+1}} \eta(s,i)ds + \int_{t_j}^{t} \eta(s,j)ds$$

and the $\mathcal{L}_p$-*norm* of η as

$$\|\eta\|_p := \lim_{t\to\infty} \left[\int_0^t |\eta(s)|^p \right]^{1/p}.$$

If $\|\eta\|_p < \infty$, then we say that $\eta \in \mathcal{L}_p$. It is worth pointing out that there holds

$$\|\eta\|_p = \|\bar{\eta}\|_p.$$

The $\mathcal{L}_\infty$-norm of η is defined as

$$\|\eta\|_\infty := \underset{(t,j)\in\text{dom}(\eta)}{\text{ess.sup}} |\eta(t,j)|$$

and we say that $\eta \in \mathcal{L}_\infty$ if $\|\eta\|_\infty < \infty$.

Consider the general reset system with time-regularization

$$\begin{cases} \left.\begin{aligned} \dot{\tau} &= 1 \\ \dot{x} &= f(x,d) \end{aligned}\right\} & \text{if } x \in \mathcal{F} \text{ or } \tau \leq \rho \\ \left.\begin{aligned} \tau^+ &= 0 \\ x^+ &= g(x) \end{aligned}\right\} & \text{if } x \in \mathcal{J} \text{ and } \tau \geq \rho \end{cases} \tag{7.18}$$

with $x \in \mathbb{R}^n$, $d \in \mathbb{R}^m$, and $\tau \in \mathbb{R}_{\geq 0}$. $\rho > 0$ is the time-regularization constant used to bound the reset intervals from below. A hybrid arc $\xi := (x, \tau)$ and a hybrid input d form a solution pair (ξ, d) for the system (7.18) if the following conditions hold:

1. $\text{dom}(\xi) = \text{dom}(d)$.
2. For all j and almost all t such that $(t,j) \in \text{dom}(\xi)$, there hold
   a. $x(t,j) \in \mathcal{F}$ or $\tau(t,j) \leq \rho$;
   b. $\dot{x}(t,j) = f(x(t,j), d(t,j))$; $\dot{\tau}(t,j) = 1$.
3. For all $(t,j) \in \text{dom}(\xi)$ such that $(t,j+1) \in \text{dom}(\xi)$, there hold
   a. $x(t,j) \in \mathcal{J}$ and $\tau(t,j) \geq \rho$;
   b. $x(t,j+1) = g(x(t,j))$; $\tau(t,j+1) = 0$.

For any given $p \in [1, \infty)$, system (7.18) is said to be *finite $\mathcal{L}_p$ gain stable* from d to x if there exist constants γ_p, $\gamma_0 > 0$ such that for any initial condition (x_0, τ_0) and any $d \in \mathcal{L}_p$, there holds

$$\|x\|_p \leq \gamma_0|x_0| + \gamma_p\|d\|_p. \tag{7.19}$$

System (7.18) is said to be *exponential input-to-state stable* from d to x, if there exist positive constants γ_∞, m, $l > 0$ such that for any initial condition (x_0, τ_0) and any $d \in \mathcal{L}_\infty$, all solutions satisfy

$$|x(t,j)| \le \max\{m|x_0|e^{-lt}, \gamma_\infty \|d\|_\infty\}, \quad \forall (t,j) \in \mathrm{dom}\,(x).$$

System (7.18) is said to be *finite gain $\mathcal{L}_p$ to $\mathcal{L}_\infty$ stable* from d to x if there exist constants $\gamma_{p,\infty}$, $\gamma_0 > 0$ such that for any initial condition (x_0, τ_0) and any $d \in \mathcal{L}_p$, there holds

$$|x(t,j)| \le \gamma_0|x_0| + \gamma_{p,\infty}\|d\|_p, \quad \forall (t,j) \in \mathrm{dom}\,(x). \tag{7.20}$$

7.2.1 Passification via reset

Consider a reset controller

$$\begin{cases} \left.\begin{aligned} \dot{\tau} &= 1 \\ \dot{x}_c &= f(x_c, v, \theta) \end{aligned}\right\} & \text{if } (u,v) \in \mathcal{F} \text{ or } \tau \le \rho \\ \left.\begin{aligned} \tau^+ &= 0 \\ x_c^+ &= 0 \end{aligned}\right\} & \text{if } (u,v) \in \mathcal{J} \text{ and } \tau \ge \rho \\ u = h(x_c, \theta) & \end{cases} \tag{7.21}$$

with $x \in \mathbb{R}^n$, u, $v \in \mathbb{R}^q$, $\theta \in \mathbb{R}^m$, and

$$\mathcal{F} = \left\{ \begin{bmatrix} u \\ v \end{bmatrix} \middle| \phi(u,v) \ge 0 \right\}, \tag{7.22}$$

$$\mathcal{J} = \left\{ \begin{bmatrix} u \\ v \end{bmatrix} \middle| \phi(u,v) \le 0 \right\} \tag{7.23}$$

where

$$\phi(u,v) = (u + \epsilon_1 v)^T (v - \epsilon_2 u). \tag{7.24}$$

The baseline vector field $f(x_c, v, \theta)$ and the output function $h(x_c, \theta)$ are assumed to satisfy the following assumption.

Assumption 7.1. *The functions f and h are continuous in their arguments and there are constants $L_f > 0$ and $L_h > 0$ such that*

$$|f(x_c, v, \theta)| \le L_f(|x_c| + |v|), \quad \forall x_c, \forall v, \forall \theta \tag{7.25}$$

$$|h(x_c, \theta)| \le L_h|x_c|, \quad \forall x_c, \forall \theta. \tag{7.26}$$

Remark 7.1. *The above assumption together with the time-regularization assures that all the solutions are non-Zeno. In addition, for any integrable input $\bar{v}$ and each measurable input $\bar{\theta}$, each solution pair $(\xi, (v, \theta))$ is complete, where v and θ are the hybrid signals lifted from $\bar{v}$ and $\bar{\theta}$ on* dom (ξ)*, respectively. See Reference 5 for details.*

The flow set and jump set for planar case is intuitively depicted in Figure 7.1. The shaped area is defined by $\phi(u,v) \ge 0$ which degenerates to $u^T v \ge 0$ when $\epsilon_1 = \epsilon_2 = 0$.

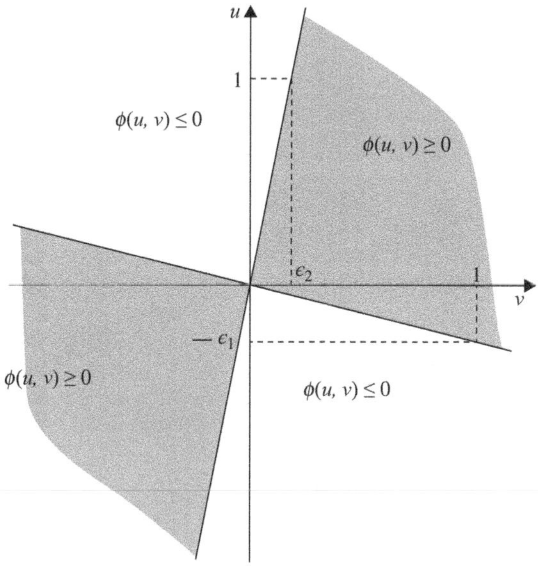

Figure 7.1 The conic subsets $\mathcal{F}$ and $\mathcal{J}$

According to the analysis in Section 7.1, if we do not introduce time-regularization, i.e., $\rho \to 0$, then the area $\mathcal{J}$ which is defined by $\phi(u, v) < 0$ is removed from the input/output space and $(v(t), u(t))$ evolves within the first and the third quadrants. By $\phi(u, v) \geq 0$, we have

$$u^T v \geq -\varepsilon_1 v^T v + \varepsilon_2 u^T u, \tag{7.27}$$

where

$$\varepsilon_1 = \frac{\epsilon_1}{1 - \epsilon_1 \epsilon_2}, \quad \varepsilon_2 = \frac{\epsilon_2}{1 - \epsilon_1 \epsilon_2}. \tag{7.28}$$

Integrate (7.27) from 0 to ∞ implies (assuming the integrals exist)

$$\int_0^\infty u^T v ds \geq \beta(x_c(0)) - \varepsilon_1 \|v\|_2^2 + \varepsilon_2 \|u\|_2^2 \tag{7.29}$$

where $\beta(x_c(0))$ is a number depending on the initial state. This means that the reset controller will have certain passivity property. Especially, since $\varepsilon_2 > 0$, it is strictly output passive.

Due to time-regularization, the actual trajectory would slightly flow into the jump set. The time-regularization constant ρ, the vector field f, and the output function h together determine how far the trajectories would go into the jump set. By Assumption 7.1, the f and h are bounded by linear functions, thus there exist constants $\tilde{\epsilon}_1 \geq \epsilon_1$ and $\tilde{\epsilon}_2 \leq \epsilon_2$ such that the flow set in the input/output space can be covered by

$$\tilde{\mathcal{F}} := \left\{ \begin{bmatrix} u \\ v \end{bmatrix} \middle| \tilde{\phi}(u, v) \geq 0 \right\}$$

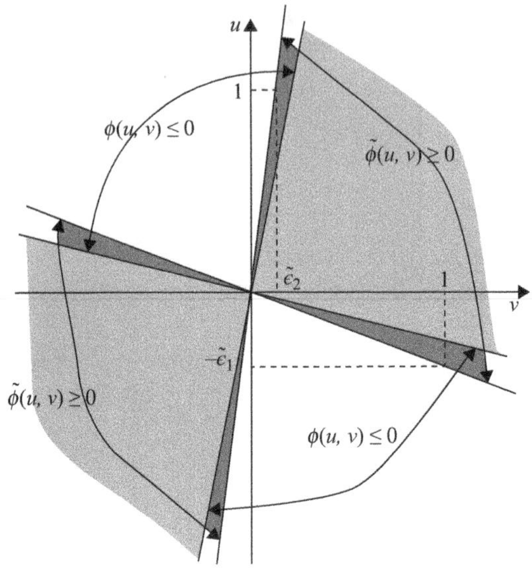

Figure 7.2 The conic subsets $\tilde{\mathcal{F}}$ and $\mathcal{J}$ when time-regularization applies

where

$$\tilde{\phi}(u, v) := (u + \tilde{\epsilon}_1 v)^T (v - \tilde{\epsilon}_2 u).$$

See Figure 7.2 for an intuitive explanation. Thus we can expect a passivity property of the following form

$$\int_0^{\infty} u^T v ds \ge \tilde{\beta} - \tilde{\varepsilon}_1 \|v\|_2^2 + \tilde{\varepsilon}_2 \|u\|_2^2 \tag{7.30}$$

where $\tilde{\varepsilon}_1$ and $\tilde{\varepsilon}_2$ should be related to ρ, L_f, and L_h, and $\tilde{\beta}$ should be related to ρ, L_f, L_h, and the initial condition $x_r(0)$. The following result gives estimates for these constants.

Theorem 7.1. *[5] Suppose that Assumption 7.1 holds. Then for a given integrable input signal $\bar{v} \in \mathcal{L}_2$, a measurable signal $\bar{\theta}$ and a solution pair $(\xi, (v, \theta))$ where $\xi = (x, \tau)$ and v and θ are hybrid signals lifted from $\bar{v}$ and $\bar{\theta}$, respectively, there holds*

$$\int_0^{\infty} \bar{u}^T \bar{v} ds \ge -\frac{2\bar{k}_\rho}{\rho L_f^2} |x_r(0)|^2 - (\varepsilon_1 + 2\bar{k}_\rho)\|\bar{v}\|_2^2 + \varepsilon_2 \|\bar{u}\|_2^2 \tag{7.31}$$

where the output signal $\bar{u}$ is projected from the hybrid output signal u : dom $(u) \to \mathbb{R}^q$ corresponding to the solution pair $(\xi, (v, \theta))$, ε_1 and ε_2 are defined in (7.28) and

$$\bar{k}_\rho := k_\rho(1 + \varepsilon_2 k_\rho), \quad k_\rho = \rho L_f L_h \max\{1, \rho e^{L_f \rho}\}. \tag{7.32}$$

Theorem 7.1 shows that the resulting reset controller is strictly output passive since $\epsilon_2 > 0$. However, it is not strictly input passive in general because

$$-(\varepsilon_1 + 2\bar{k}_\rho) \le 0.$$

This situation is due to the fact that $\epsilon_1 > 0$ which makes sure that the subset $\{(u, v) \mid u = 0, v \neq 0\}$ is part of the interior of $\mathcal{F}$. See Figure 7.1. Thus, after each reset, the state of the controller belongs to the interior of the flow set. However, if this kind of reset controller is interconnected with a plant via negative feedback, this shortage of input passivity could be compensated for if the plant is strictly output passive. Another way to make the reset controller very strictly passive is to introduce a feedforward [5, 6] as

$$\begin{cases} \left.\begin{aligned} \dot{\tau} &= 1 \\ \dot{x}_c &= f(x_c, v, \theta) \end{aligned}\right\} & \text{if } (u, v) \in \mathcal{F} \text{ or } \tau \le \rho \\ \left.\begin{aligned} \tau^+ &= 0 \\ x_c^+ &= 0 \end{aligned}\right\} & \text{if } (u, v) \in \mathcal{J} \text{ and } \tau \ge \rho \\ u = h(x_c, \theta) + \epsilon_3 v \end{cases} \tag{7.33}$$

where $\mathcal{F}$ and $\mathcal{J}$ are defined in (7.22) and (7.23), respectively, with

$$\phi(u, v) := [u + (\epsilon_1 - \epsilon_3)v]^T[(1 + \epsilon_2\epsilon_3)v - \epsilon_3 u]. \tag{7.34}$$

Define

$$\tilde{u} = u - \epsilon_2 v, \tag{7.35}$$

then $\tilde{u} = h(x_c, \theta)$ and

$$\phi(u, v) = \tilde{\phi}(\tilde{u}, v) = (\tilde{u} + \epsilon_1 v)^T(v - \epsilon_2\tilde{u}). \tag{7.36}$$

By using Theorem 7.1 (viewing $\tilde{u}$ as the output), there holds

$$\int_0^\infty \bar{\tilde{u}}^T \bar{v} ds \ge -\frac{2\bar{k}_\rho}{\rho L_f^2}|x_r(0)|^2 - (\varepsilon_1 + 2\bar{k}_\rho)\|\bar{v}\|_2^2 + \varepsilon_2\|\bar{\tilde{u}}\|_2^2 \tag{7.37}$$

where $\bar{\tilde{u}}$ is the continuous-time signal projected from the hybrid signal $\tilde{u}$ on $\mathbb{R}_{\ge 0}$. Substituting (7.35) into the above inequality leads to

$$\int_0^\infty \bar{u}^T \bar{v} ds \ge -\beta + \eta_1\|\bar{u}\|_2^2 + \eta_2\|\bar{v}\|_2^2 \tag{7.38}$$

where

$$\begin{aligned} \eta_1 &= \frac{\varepsilon_2}{1 + 2\varepsilon_2\epsilon_3}, \quad \eta_2 = \frac{\epsilon_3 + \varepsilon_2\epsilon_3^2 - \varepsilon_1 - 2\bar{k}_\rho}{1 + 2\varepsilon_2\epsilon_3}, \\ \beta &= \frac{2\bar{k}_\rho}{(1 + 2\varepsilon_2\epsilon_3)\rho L_f^2}|x_r(0)|^2. \end{aligned} \tag{7.39}$$

Thus, if we choose a ϵ_3 such that

$$\epsilon_3 + \varepsilon_2\epsilon_3^2 > \varepsilon_1 + 2\bar{k}_\rho,$$

then the reset controller is very strictly passive.

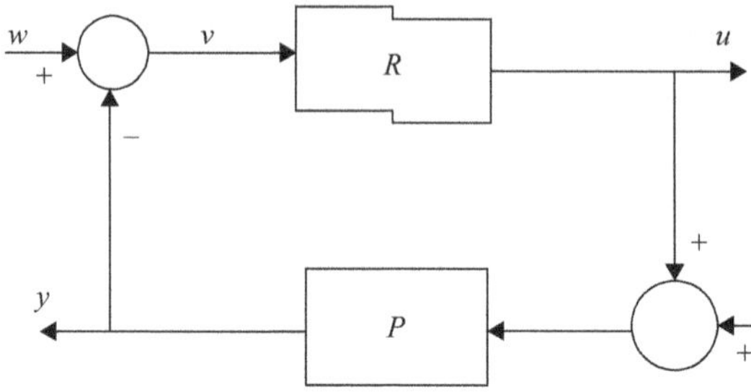

Figure 7.3 Negative feedback interconnection of reset controller with a plant

7.2.2 Finite $\mathcal{L}_2$-gain stability

Consider the interconnection of a reset controller with the plant

$$\begin{cases} \dot{x}_p = f_p(x_p, u+d) \\ y = h_p(x_p, u+d) \end{cases} \tag{7.40}$$

via negative feedback

$$v = w - y. \tag{7.41}$$

By using the passivity properties of the reset controllers in the previous section and Passivity Theorem (Theorem 3.4), we have the following result.

Theorem 7.2. *[5] Consider the feedback interconnection depicted in Figure 7.3, where the plant (P) is given by (7.40) and the reset controller (R) is given by (7.21), (7.22), (7.23), and (7.24) satisfying Assumption 7.1. For any* $\epsilon_1 \geq 0$, $\epsilon_2 > 0$, *and* $\rho > 0$, *if the plant is output strictly passive with an excess of output passivity*

$$\delta_p > \varepsilon_1 + 2\bar{k}_\rho$$

where ε_1 *and* $\bar{k}_\rho$ *are defined in (7.28) and (7.32), respectively, then the closed-loop system is finite* $\mathcal{L}_2$*-gain stable from* (w, d) *to* (u, v).

By the definition of ε_1 and $\bar{k}_\rho$, there hold

$$\lim_{\epsilon_1 \to 0} \varepsilon_1 = 0, \quad \lim_{\rho \to 0} \bar{k}_\rho = 0.$$

This leads to the following result.

Theorem 7.3. *[5] Consider the feedback interconnection depicted in Figure 7.3, where the plant (P) is given by (7.40) which is output strictly passive, and the reset controller (R) is given by (7.21), (7.22), (7.23), and (7.24) satisfying Assumption 7.1. Then for any* $\epsilon_2 > 0$, *there exist small enough positive numbers* ϵ_1^* *and* ρ^* *such that for all* $\epsilon_1 \in (0, \epsilon_1^*]$ *and all* $\rho \in (0, \rho^*]$, *the closed-loop system is finite* $\mathcal{L}_2$*-gain stable from* (w, d) *to* (u, v).

Theorem 7.4. *[5] Consider the feedback interconnection depicted in Figure 7.3, where the plant (P) is given by (7.40) which is passive, and the reset controller (R) is given by (7.33), (7.22), (7.23), and (7.34) satisfying Assumption 7.1. For any* $\epsilon_1 \geq 0$, $\epsilon_2 > 0$, *and* $\rho > 0$, *if*

$$\epsilon_3 + \varepsilon_2\epsilon_3^2 > \varepsilon_1 + 2\bar{k}_\rho,$$

then the closed-loop system is finite $\mathcal{L}_2$*-gain stable from* (w, d) *to* (u, v).

The above results are based on the passivity theorem. For the general reset system with conic jump set described in (7.18), we have the following Lyapunov-function-based result.

Theorem 7.5. *[4] Suppose that* $f(\cdot,\cdot)$ *is continuous in its first argument and the output function satisfies* $|g(x)| \leq \lambda_0|x|$. *For any given* $p \in [1, \infty)$, *if there is a function* $V(x)$, *positive numbers* λ_i, $i = 1, \ldots, 7$, *such that for all d,*

$$\lambda_1|x|^p \leq V(x) \leq \lambda_2|x|^p, \tag{7.42}$$

$$\frac{\partial V(x)}{\partial x} f(x,d) \leq \lambda_3 V(x) + \lambda_4|x|^{p-1}|d|, \quad \forall\, a.a\; x \in \mathbb{R}^n, \tag{7.43}$$

$$\max_{v\in\partial V(x)} v^T f(x,d) \leq -\lambda_5 V(x) - \lambda_6|y|^p + \lambda_7|d|^p, \quad x \in \mathcal{F}, \tag{7.44}$$

$$V(g(x)) \leq \eta V(x), \quad \forall x \in \mathcal{J}, \tag{7.45}$$

where $\partial V(x)$ *is the Clarke generalized gradient of* $V(x)$. *Then, we have the following:*

1. *There exists a* $\rho^* > 0$ *such that for any* $\rho \in (0, \rho^*]$ *the origin of the x dynamics of (7.18) with* $d = 0$ *is exponentially stable;*
2. *The system (7.18) is finite* $\mathcal{L}_p$ *gain stable from d to y, and for any* $\epsilon > 0$, *there exists* ρ^* *such that for all* $\rho \leq \rho^*$, *the* $\mathcal{L}_p$ *gain from d to y is upper-bounded by*

$$\left[\frac{\lambda_7}{\lambda_6}\right]^{1/p} + \epsilon.$$

Theorem 7.6. *[4] Consider RCS (7.18) with*

$$f(x,d) = Ax + Bd, \tag{7.46}$$

$$g(x) = A_r x, \tag{7.47}$$

where $A, A_r \in \mathbb{R}^{n\times n}$, $B \in \mathbb{R}^{n\times m}$. *Suppose that the set* $\mathcal{J}$ *is a closed nonempty cone, then the following statements are equivalent:*

1. *The origin of the x dynamics of (7.18) with* $d = 0$ *is (locally) asymptotically stable;*
2. *The origin of the x dynamics of (7.18) with* $d = 0$ *is globally exponentially stable;*

3. *Given $p \in [1,\infty)$, system (7.18) is finite $\mathcal{L}_p$ gain stable and $\mathcal{L}_p$ to $\mathcal{L}_\infty$ stable from d to x;*
4. *System (7.18) is finite gain exponentially input-to-state stable from d to x;*
5. *Given $p \in [1,\infty)$, there exists a function $W(t,x) \geq 0$ that is continuously differentiable in $\mathbb{R} \times \mathbb{R}^n \setminus \{0\}$ and positive constants λ_i, $i = 1,\ldots,8$, and $\nu \in (0,1)$, such that for all d the following bounds are satisfied*

$$\lambda_1|x|^p \leq W(\tau,x) \leq \lambda_2|x|^p, \quad \forall(\tau,x) \in \mathbb{R} \times \mathbb{R}^n, \tag{7.48}$$

$$\frac{\partial W(\tau,x)}{\partial \tau} + \left[\frac{\partial W(\tau,x)}{\partial x}\right]^T (Ax + Bd),$$
$$\leq -\lambda_5 W(\tau,x) - \lambda_6|x|^p + \lambda_7|x|^{p-1}|d|, \quad \forall(\tau,x) \in \bar{\mathcal{F}} \setminus \{0\}, \tag{7.49}$$

$$W(0,A_r x) \leq \nu W(\tau,x), \quad \forall(\tau,x) \in \bar{\mathcal{J}}, \tag{7.50}$$

$$\left|\frac{\partial W(\tau,x)}{\partial x}\right| \leq \lambda_8|x|^{p-1}, \quad \forall(\tau,x) \in \bar{\mathcal{F}} \setminus \{0\}, \tag{7.51}$$

where

$$\bar{\mathcal{F}} := \{(\tau,x) \mid (x \in \mathcal{F} \text{ and } \tau \in [\rho,\rho+1]) \text{ or } \tau \in [0,\rho]\},$$

$$\bar{\mathcal{J}} := \{(\tau,x) \mid (x \in \mathcal{J} \text{ or } \tau \in (\rho,\rho+1]) \text{ or } \tau \in (-\infty,0] \cup [\rho+1,+\infty)\}.$$

Notes

The materials of this chapter mainly come from References 5 and 4. Only the main idea of RCSs with conic jump sets and part of the results regarding passification via reset and finite $\mathcal{L}_2$-gain stability are included in this chapter. Most of the proofs of the theorems are omitted which can be found in the literature. An upper bound estimate of the $\mathcal{L}_2$ gain for planar RCSs can be found in Reference 4. The reader is referred to Reference 7 for some other results on finite $\mathcal{L}_2$-gain stability for RCSs with conic jump sets. A comprehensive discussion of hybrid systems can be found in Reference 2.

References

[1] L. Zaccarian, D. Nesic, and A.R. Teel. First order reset elements and the Clegg integrator revisited. *American Control Conference, 2005*, June 8–10, 2005, vol. 1, Portland, OR, USA, pp. 563–568.

[2] R. Goebel, R.G. Sanfelice, and A.R. Teel. *Hybrid dynamical systems: modeling, stability, and robustness*, Princeton University Press, Princeton, NJ, 2012.

[3] S. Boyd, L. El Ghaoul, E. Feron, and V. Balakrishnan. Linear matrix inequalities in System and Control Theory. In *SIAM Studies in Applied Mathematics*, vol. 15,

Society for Industrial and Applied Mathematics, Philadelphia, Pennsylvania, 1994.

[4] D. Nešic, A. Teel, and L. Zaccarian. Stability and performance of SISO control systems with first order reset elements. *IEEE Transactions on Automatic Control*, 56(11):2567–2582, 2011.

[5] F. Forni, D. Nešić, and L. Zaccarian. Reset passivation of nonlinear controllers via a suitable time-regular reset map. *Automatica*, 47(9):2099–2106, 2011.

[6] H.K. Khalil. Nonlinear systems. Prentice Hall, Inc., Upper Saddle River, NJ, 1996.

[7] D. Nešic, L. Zaccarian, and A.R. Teel. Stability properties of reset systems. *Automatica*, 44(8):2019–2026, 2008.

Index